AF347535

Advances and Trends in Mathematical Modelling, Control and Identification of Vibrating Systems

Advances and Trends in Mathematical Modelling, Control and Identification of Vibrating Systems

Editors

Francisco Beltran-Carbajal
Julio Cesar Rosas-Caro
Juan M Ramirez
Roberto Salvador Félix Patrón

MDPI • Basel • Beijing • Wuhan • Barcelona • Belgrade • Manchester • Tokyo • Cluj • Tianjin

Editors

Francisco Beltran-Carbajal
Departamento de Energía,
Universidad Autónoma
Metropolitana, Unidad
Azcapotzalco
Mexico

Julio Cesar Rosas-Caro
Faculty of Engineering,
Universidad Panamericana
Sede Guadalajara
Mexico

Juan M Ramirez
Department of Electrical
Engineering (Guadalajara),
Center for Research and
Advanced Studies of the
National Polytechnic Institute
Mexico

Roberto Salvador Félix Patrón
Department of Aviation
Academy, Centre for Applied
Research on Education,
Amsterdam University of
Applied Sciences
The Netherlands

Editorial Office
MDPI
St. Alban-Anlage 66
4052 Basel, Switzerland

This is a reprint of articles from the Special Issue published online in the open access journal *Mathematics* (ISSN 2227-7390) (available at: https://www.mdpi.com/journal/mathematics/special_issues/Mathematical_Modelling_Control_Identification_Vibrating_Systems).

For citation purposes, cite each article independently as indicated on the article page online and as indicated below:

LastName, A.A.; LastName, B.B.; LastName, C.C. Article Title. *Journal Name* **Year**, *Volume Number*, Page Range.

ISBN 978-3-0365-3949-2 (Hbk)
ISBN 978-3-0365-3950-8 (PDF)

Contents

About the Editors

Francisco Beltran-Carbajal received his B.S. in Electromechanical Engineering from the Instituto Tecnológico de Zacatepec (Mexico) and his Ph.D. in Electrical Engineering (Mechatronics) from the Centro de Investigacion y Estudios Avanzados del Instituto Politecnico Nacional (CINVESTAV-IPN) in Mexico City. He is currently a Titular Professor in the Energy Department at Universidad Autonoma Metropolitana (UAM), Unidad Azcapotzalco in Mexico City. His main research interests are vibration control, system identification, rotating machinery, mechatronics, and automatic control of energy conversion systems.

Julio C. Rosas-Caro received a B.S. in electronics and an M.S. in E.E. from the Tecnologico de Cd. Madero, Mexico, in 2004 and 2005, respectively, and a Ph.D. in E.E. from the Cinvestav del IPN, Mexico, in 2009. He has been a Visiting Scholar at Michigan State University, the University of Colorado Denver, and the Ontario Tech. Dr. Rosas-Caro is an associate editor of the IEEE Latin America Transactions. He is currently a Professor at Universidad Panamericana, Mexico. His research interest is power electronics, including dc–dc converters, flexible alternating current transmission system devices, power converter topologies, and applications.

Juan M Ramirez obtained his Ph.D. in Mexico in 1992. Since 1999, he has been with the Centro de Investigacion y de Estudios Avanzados del IPN, Unidad Guadalajara, Mexico. His areas of interest are smart grids, optimization, dynamic analysis of electrical networks and power electronics applications. Since 1990, he has been a member of the Mexican National System of Researchers.

Roberto Félix Patrón obtained a double degree in Electronics Engineering at the Universidad Autónoma de Baja California in Tijuana, Mexico, and the ENSEEIHT in Toulouse, France. He followed with a master's degree in Computer Science, Automation and Decision-Making at the Institut National Polytechnique de Toulouse, and a Ph.D. in Aerospace Engineering at the École de Technologie Supérieure in Montreal, Canada. Roberto has experience in the design and implementation of mathematical optimization models to aircraft performance databases, and he is currently a researcher in data science topics to improve aviation operations. He has worked as an aircraft systems simulation engineer and he has participated in projects related to motor control applications in the past.

Preface to "Advances and Trends in Mathematical Modelling, Control and Identification of Vibrating Systems"

This book introduces novel results in mathematical modelling, parameter identification, and efficient automatic control for a wide range of applications of mechanical, electric, and mechatronic systems, where undesirable oscillations or vibrations are manifested. The six chapters of the book written by experts from the international scientific community cover a wide range of interesting research topics related to original and innovative contributions to identification techniques of rotordynamic parameters in rotor-bearing systems, finite element modelling, active vehicle suspension systems, model-free data-driven-based control, voltage source converters, static synchronous compensators, bending vibrations in flexible structures, active vibration control on quadrotor aerial vehicles, artificial neural networks, particle swarm optimization, and low-frequency oscillations in large-scale power systems. The book is addressed to both academic and industrial researchers and practitioners, as well as to postgraduate and undergraduate engineering students and other experts in a wide variety of disciplines seeking to know more about the advances and trends in mathematical modelling, control, and identification of engineering systems in which undesirable oscillations or vibrations could be presented during their operation.

The book is organized into six chapters. A brief description of every chapter follows. Chapter 1 deals with the problem that arises when diverse regulation devices and controlling strategies are involved in electric power systems' regulation design. A B-Spline neural networks algorithm is used to define the best controllers gains to efficiently attenuate low-frequency oscillations when a short circuit event is presented. Chapter 2 introduces an exact elastodynamics theory for bending vibrations for a class of flexible structures, which is based on the partial differential operator theory. In Chapter 3, the authors describe a model-free data-driven-based control for a Voltage Source Converter (VSC)-based Static Synchronous Compensator (STATCOM) to improve the dynamic power grid performance under transient scenarios. Chapter 4 proposes a planned motion profile tracking control scheme and vibrating disturbance suppression for quadrotor aerial vehicles using artificial neural networks and particle swarm optimization. Chapter 5 presents a model predictive control method for active automotive suspension systems by means of hydraulic actuators. Chapter 6 concludes the book, describing fast algebraic identification techniques of rotordynamic parameters in rotor-bearing systems using finite element models.

Finally, we would like to express our sincere gratitude to all the authors for their excellent contributions, which we are sure will be valuable to the readers. We hope that this book can be useful and inspiring for contributing to the technology development, new academic and industrial research, and many inventions and innovations in the field of mathematical modelling, control, and identification of mechanical, electric, and mechatronic systems where vibrations or oscillations could be exhibited.

Dr. Francisco Beltran-Carbajal

Universidad Autónoma Metropolitana, Unidad Azcapotzalco, Departamento de Energía

Dr. Julio Cesar Rosas-Caro

Universidad Panamericana, Sede Guadalajara, Facultad de Ingeniería, Mexico

Dr. Juan M Ramirez

Centro de Investigacion y de Estudios Avanzados del IPN, Unidad Guadalajara, Mexico.

Dr. Roberto Félix Patrón

Department of Aviation Academy, Centre for Applied Research on Education, Amsterdam University of Applied Sciences

Francisco Beltran-Carbajal, Julio Cesar Rosas-Caro, Juan M Ramirez, and Roberto Salvador Félix Patrón

Editors

mathematics

Article

A Novel Methodology for Adaptive Coordination of Multiple Controllers in Electrical Grids

Ruben Tapia-Olvera [1], Francisco Beltran-Carbajal [2], Antonio Valderrabano-Gonzalez [3,*] and Omar Aguilar-Mejia [4]

[1] Department of Electrical Energy, Universidad Nacional Autónoma de México, Mexico City 04510, Mexico; rtapia@fi-b.unam.mx
[2] Department of Energy, Universidad Autónoma Metropolitana, Unidad Azcapotzalco, Mexico City 02200, Mexico; fbeltran@azc.uam.mx
[3] Facultad de Ingeniería, Universidad Panamericana, Álvaro del Portillo 49, Zapopan 45010, Mexico
[4] School Engineering, UPAEP University, Puebla 72410, Mexico; omar.aguilar@upaep.mx
* Correspondence: avalder@up.edu.mx

Abstract: This proposal is aimed to overcome the problem that arises when diverse regulation devices and controlling strategies are involved in electric power systems regulation design. When new devices are included in electric power system after the topology and regulation goals were defined, a new design stage is generally needed to obtain the desired outputs. Moreover, if the initial design is based on a linearized model around an equilibrium point, the new conditions might degrade the whole performance of the system. Our proposal demonstrates that the power system performance can be guaranteed with one design stage when an adequate adaptive scheme is updating some critic controllers' gains. For large-scale power systems, this feature is illustrated with the use of time domain simulations, showing the dynamic behavior of the significant variables. The transient response is enhanced in terms of maximum overshoot and settling time. This is demonstrated using the deviation between the behavior of some important variables with StatCom, but without or with PSS. A B-Spline neural networks algorithm is used to define the best controllers' gains to efficiently attenuate low frequency oscillations when a short circuit event is presented. This strategy avoids the parameters and power system model dependency; only a dataset of typical variable measurements is required to achieve the expected behavior. The inclusion of PSS and StatCom with positive interaction, enhances the dynamic performance of the system while illustrating the ability of the strategy in adding different controllers in only one design stage.

Keywords: B-spline neural networks; adaptive power system control; coordinated multiple controllers; StatCom

Citation: Tapia-Olvera, R.; Beltran-Carbajal, F.; Valderrabano-Gonzalez, A.; Aguilar-Mejia, O. A Novel Methodology for Adaptive Coordination of Multiple Controllers in Electrical Grids. *Mathematics* **2021**, *9*, 1474. https://doi.org/10.3390/math9131474

Academic Editor: Cristina I. Muresan

Received: 19 May 2021
Accepted: 15 June 2021
Published: 23 June 2021

1. Introduction

Electric power systems are large, interconnected, complex, and highly changeable systems that are always affected by a wide variety of perturbations [1]. Therefore, the control design stage and tuning procedure for multiple controllers is an entangled task [2,3], present interesting approaches on stabilizing procedures in electric power systems that use multiple power system stabilizers with lead and lag compensators. The conventional linear controllers designed around an equilibrium point are useful, but their performance could be degraded if variations are presented in the system. On the other hand, dealing with non-linear controllers is a high demanding and slow task due to the complexity of large-scale power system. In general, for reaching a good performance, these strategies present dependency on the parameters system modeling.

Power system stabilizers (PSS) have been used to generate supplementary signals to control the excitation system to improve the power system dynamic performance by the damping of system oscillations [1]. However, the expected behavior depends entirely on the correct selection of controllers' gains and time constants [2,3]. Moreover, some flexible alternating current transmission systems (FACTS devices) are included to solve some specific power systems problems; nevertheless, their operation is also depending upon the positive interaction with other regulation devices. Refs. [4,5] exemplify the problem of simultaneous tuning of multiple controllers in large scale power system including FACTS devices in transmission systems.

There are several methodologies to solve the problem of designing linear controllers to reach good dynamic performance. However, these solutions are complex in implementation; they do not cover a wide range of operating conditions of the power system or they do not have the same behavior with new grid topologies. The main objective of this proposal is to attain an adaptive performance of PSS in large-scale power systems with the possibility of adding new components that change the grid configuration, in this case for exemplifying through a static synchronous compensator (StatCom).

In order to validate the proposed strategy and without loss of generality, this paper presents the control design problem of PSS in power systems including a StatCom, which is one of the most useful FACTS devices in practical power systems. This configuration adds enough complexity to verify the viability of the proposal.

In general, the design control stage has been considered an independent problem, with only one controller. The fact that the system can have other regulation devices, has not been included. Only few works contemplate more than one controller simultaneously in the design stage. However, this is an open research topic due to the electrical grid composition and the continuous topology changing on it [6].

In [2] two objective functions must be solved to obtain coordination between PSS and traditional static VAR compensators (SVC). In order to reduce the high computational load, the genetic algorithm was used for solving the multi-objective optimization problem, adapting it for parallel computing. An analysis based on the power system modeled as a set of hybrid non-linear differential algebraic equations is presented in [3], where the dynamic behavior of the system is studied in various scenarios: no PSS, PSS without dead-band, and PSS with dead-band.

In [7], a single machine infinite bus (SMIB) model is used to tune the PSS, and then new non-specified adjustments are carried out to extend the scheme to the multimachine scenario. Additionally, the tuning stage is very case-dependent. Multi-band PSS are tuned in [8] by using an optimization search method based on modal performance index, but representative linearized system models are required for the optimization procedure.

The PSS tuning based on linear quadratic regulator design is presented in [9]. The state and input matrices of the linearized power system model are required for developing the optimization procedure in a single machine case, and then it is extended to the multimachine case. In [10], a two-level control strategy that blends a local controller with a centralized controller is proposed to diminish low frequency oscillations. In the PSS model, a proportional integral (PI) controller is added. Two extra gains are included in the problem solution. For the tuning procedure, two stages are required, first the design of the local PI controller and then the design of the centralized controller.

A design method using a modified Nyquist diagram with an embedded partial pole-placement capability is presented in [11]. The small signal stability model obtained by the linearization of the power system around an operating point is required. That method evaluates the open loop transfer function along a line of constant damping ratio to design PSS for two test systems.

Additionally, control design based on non-linear theory is used, but in the same sense the procedure is realized separately for each controller. In [12], a scheme called decentralized continuous higher-order sliding mode excitation control is applied. The deviations on the angle of the power are required to obtain the desired system performance, also the estimation of first and second order time derivatives of this angle must be determined. Similarly, in [13] the H_∞ control with regional pole placement is used to ensure adequate power system dynamic performance, the linearized model around an equilibrium point is also needed. Additionally, deterministic strategies based on artificial intelligence could be an alternative to the design procedure of multiple controllers in electrical grids [14]. Another important algorithm is the non-linear feed forward control which represents an option of non-linear adaptive control techniques [15]. This kind of strategies has been little explored in applications for electrical power systems. Similarly, other approach that can be extended to large scale power systems is the physics-based control technique [16].

A scheme called networked predictive control (NPC) used to design a damping controller that incorporates a generalized predictive control (GPC) to generate optimal control predictions is presented in [17]. Model identification is required to deal with uncertainties and to provide an adaptive predictive model for GPC. This method describes four steps for designing a NPC for a wide area damping controller: (i) modal analysis of the detailed non-linear model; (ii) determination of the order of the reduced order model of the power system; (iii) obtain the low-order equivalent model via model identification algorithm and use it as the prediction model for the NPC; (iv) selection of parameters like the output prediction horizon, the control horizon, the weighting sequence, and the sampling period.

Finally, artificial intelligence methodologies such as artificial neural networks (ANN), fuzzy logic (FL), or neuro-fuzzy are used for design purposes. In [18] an adaptive fuzzy sliding mode controller with a PI switching surface to damp power system oscillations is proposed. This strategy combines: (a) a sliding surface, (b) a fuzzy controller, (c) a curbing controller, and (d) a wavelet neural network to obtain the best auxiliary signal input to the excitation system. The structure of wavelet neural network is based on three layers, where the inputs are the sliding surface and its derivative.

A so-called hybrid adaptive non-linear controller is proposed in [19]. For the controller design it is necessary to estimate non-linear parts of the system, it is also required to measure data. The controller has a feed forward neural network structure, it is trained offline with extensive test data and it is adjusted online. In [20], the design of a PSS based on a combination of fuzzy logic and sliding mode theory is illustrated. This proposal indicates that a fuzzy-PID controller is composed of fuzzy PI and fuzzy PD controllers, and the response depends on scaling factors, hence selection of these parameters is crucial while designing the controller. The definition of the fuzzy rules is also an important issue for its correct operation.

Other important proposals, including FACTS devices, offer better results working with positive interaction with PSS. In [4], an optimization formulation is used to coordinate one PSS with one unified power flow controller (UPFC), but two objective functions based on eigenvalues of the state are needed for it. The possibility of using different FACTS devices is indicated in [5], the results include a StatCom and a UPFC. The eigenvalues of the power system model are required on the tuning procedure.

In [21], a StatCom and a PSS have been tuned to get a good dynamic power system performance using the seeker optimization algorithm to obtain the controller gains by an objective function. The StatCom model used, includes the components of the current and voltage dynamic in terminals of direct current (DC) capacitor.

Similarly, an objective function in [22] is used to attain a positive interaction between StatCom and PSS with a constraint set. The StatCom model is described with the operating range curve, but no dynamic equations are included. In [23], the dynamic operation of the StatCom is coordinated with a PSS. The tuning procedure depends on an objective function, and the definition of a constrain set.

The changing nature of power systems demands different types of studies due the inclusion of new control devices, renewable energies, and emerging technologies. However, it is difficult to have a unique methodology to solve the problem of the control design in large-scale power system. Although there are different alternatives to solve this problem, these proposals offer a solution limited to the characteristics of the systems under study. In multimachine power systems the control design problem is amplified due to the presence of multiple controllers that must be tuned simultaneously to guarantee a positive interaction for each operating condition.

Therefore, the present contribution considers the non-linear power system nature and it defines an adaptive controllers' behavior. This performance is obtained by the inclusion of some selected dynamic gains that are updated on each sample time to find the best values for every operating condition and system topology. It is possible to update all the controller gains, but to exemplify the relevance of the proposal, only some of them are dynamically calculated. Simultaneous tuning of each controller is obtained.

To validate the proposed scheme based on B-Spline neural networks, PSS are simultaneously coordinated with a StatCom to enhance the power system dynamic response under severe disturbances. An effective control design procedure for power system controllers is demonstrated by the obtained results, improving the overall multimachine system dynamic performance. The proposal avoids the parameters and power system model dependency by using only measurements of some system variables to reach the expected behavior. The main contributions of our methodology are: (i) a new method for tuning multiple controllers in electrical grids is proposed; (ii) a time-domain analysis for damping low frequency oscillations considering different controllers when previous design stage was already performed is included; (iii) different controllers preserving good performance without imposing a particular requirement are considered; (iv) the introduced methodology offers a practical way to obtain adaptive behavior of controllers with simultaneously tuning, and positive interaction; (v) the proposed algorithm is learning online, which means no additional stages for training are required.

2. Electric Grid Operation and Control

Transient stability in large-scale power system is usually demonstrated by time domain simulations over a range of operating conditions and perturbations due to the complexity to dealing with large non-linear models associated to the power systems. Typically, the most demanding scenarios are first analyzed to have the power system with good dynamic performance, and then, similar or better behavior is expected when less demand occurs.

On the other hand, the classical stability analysis based on the power system linearized model has high complexity to attain an accurate linearized model, moreover, new components integration, and the consideration of continuous grid change involves new equilibrium points. These aspects represent another important open research topic.

Thus, we used a complete non-linear representation for transient stability studies in large-scale power system. Besides that, our proposal is proved under three phase faults, which are considered severe disturbances. The solution under these considerations is gotten by numerical methods involving a set of non-linear differential equations modeling all grid components with dynamic behavior.

Some models available in the literature are used to evaluate the proposed strategy. Additionally, the steady state condition and dynamic performance of the power system with excitation is developed in PSS® E. The results gotten are consistent in our simulation platform and the commercial software.

2.1. Power System Model

For transient stability studies, a synchronous generator model with four state variables δ_i, ω_i, E'_{qi}, E'_{di}, and an automatic voltage regulator represented by a state variable E_{fdi} [1,24] is used Equation (1). Where subscript i identifies the ith generator. Then,

$$\frac{d\delta_i}{dt} = \omega_i - \omega_B$$

$$\frac{d\omega_i}{dt} = \frac{\omega_B}{2H_i}\left[-D(\omega_i - \omega_B) + P_{mi} - P_{ei}\right]$$

$$\frac{dE'_{qi}}{dt} = \frac{1}{T'_{d0i}}\left[-E'_{qi} + \left(x_{di} - x'_{di}\right)i_{di} + E_{fdi}\right]$$

$$\frac{dE'_{di}}{dt} = \frac{1}{T'_{q0i}}\left[-E'_{di} - \left(x_{qi} - x'_{qi}\right)i_{qi}\right] \tag{1}$$

where δ is the load angle; ω is the angular speed; E'_q and E'_d are the quadrature and direct internal transient voltages, respectively; P_e is the injected real power; i_q and i_d are the quadrature and direct axis currents, respectively; E_{fd} is the excitation voltage; ω_B is the speed in steady state condition; H is the inertia constant; T'_{d0} and T'_{q0} are the d and q open-circuit transient time constants; x'_d and x'_q are the d and q transient reactances; x_d and x_q are the d and q synchronous reactances; D is the damping constant. Considering this representation, the real power is obtained by,

$$P_{ei} = E'_{di}i_{di} + E'_{qi}i_{qi} + \left(x'_{di} - x'_{qi}\right)i_{di}i_{qi} \tag{2}$$

This set of equations is solved along with the algebraic equations of the electric grid. The initial values of dq-axis currents are obtained by power flow analysis. The algebraic equations of the power grid are formulated by power flow representation and solved together with synchronous generator equations [24]. Additionally, a static excitation system is considered to regulate the terminal voltage in each equivalent model of synchronous generators,

$$\frac{dE_{fdi}}{dt} = \frac{1}{T_{Ai}}\left[-E_{fdi} + K_{Ai}\left(V_{refi} - V_{ti} + V_{si}\right)\right] \tag{3}$$

where V_{ref} is the reference voltage; V_t is the terminal voltage magnitude; V_s is the PSS's output signal (auxiliary signal); K_A and T_A are the system excitation gain and time constant.

The power system stabilizer model has the representation by phase lag-lead compensators and a washout block. The error between the actual speed and the corresponding in steady state condition is considered as the input signal, $\omega_i(s) - \omega_B$. This auxiliary control signal, V_s, must guarantee a faster damping of the low frequency oscillations that occur in the system after a short circuit failure is presented. For this purpose, it is necessary to define properly: K_s, T_w, T_1, T_2, T_3 y T_4, for each PSS included in the power system. In general, it is considered that $T_1 = T_3$ y $T_2 = T_4$.

In Figure 1, the proposed adaptive scheme is included in the power system stabilizer model to attain improved dynamic performance. This non-linear model is used to validate the tuning on the proposal. The Equations (1)–(3) are not used for design purposes. The time response of some variables is used to train the adaptive scheme in offline stage and then also in online learning operation.

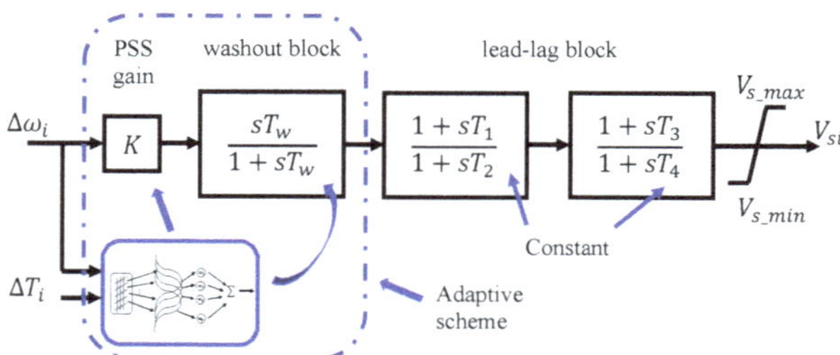

Figure 1. Improved power system stabilizers with adaptive scheme.

2.2. Statcom Model and Control

The StatCom model used on this paper consists of an equivalent transformer that emulates the voltage source converter operation. This transformer is connected in one side to a capacitor bank and, on the other side to the electric grid through a coupling transformer [25], Figure 2. One important feature of this model is the possibility to be included in transient stability studies of the power systems. The internal AC voltage of the StatCom is defined by,

$$V_{int} = km_a E_{dc} e^{j\phi} \tag{4}$$

where k is a known constant; E_{dc} is the DC voltage on the capacitor terminals; ϕ is the phase angle of V_{int} in phasor form, and; m_a in this model emulated the index modulation to regulate the voltage magnitude.

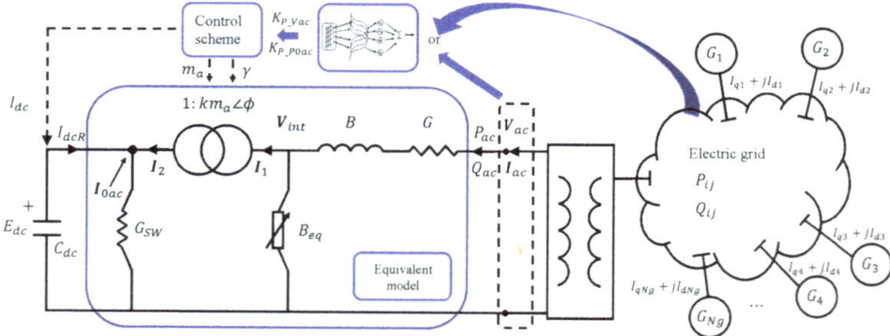

Figure 2. StatCom model with dynamic control gains.

Three PI controllers are used to regulate the StatCom dynamic performance, the main objective is to control the voltage magnitude at the point of common coupling (PCC), but an auxiliary signal could be included. In this scheme, the controlled voltage is after internal losses of the VSC, V_{ac}, before the PCC transformer. The real and reactive power of the presented equivalent circuit is defined by [25],

$$P_{ac} = V_{ac}^2 G - km_a V_{ac} E_{dc}[G\cos\gamma + B\sin\gamma]$$
$$Q_{ac} = -V_{ac}^2 B - km_a V_{ac} E_{dc}[G\sin\gamma - B\cos\gamma] \tag{5}$$

where $\gamma = \theta_{ac} - \phi$, represents the angular aperture between the internal voltage of VSC model and terminals, after internal losses, G and B. This angle is the second control variable to guarantee the desired exchange of active power, in this case only the required active power from the grid for losses compensation of the StatCom, G and G_{sw}, the last one represents the switching losses. These power flow equations are solved together with the electrical grid.

The dynamic performance is evaluated by the resulting equations of the equivalent circuit in Figure 2. In the DC bus,

$$i_c = C_{dc}\dot{E}_{dc} \tag{6}$$

where $i_c = -I_{dcR} - I_{dc}$, also,

$$I_{dcR} = \frac{P_{0ac}}{E_{dc}} \tag{7}$$

where I_{dc} is the output of the first PI controller with $E_{dc} - E_{dc}^{nom}$ as its input. Two controller gains, $K_{P_E dc}$ and $K_{I_E dc}$ are needed.

The capacitor cannot inject active power, so it is necessary one regulator to guarantee the physical condition that only active power losses are absorbed from the grid. Therefore, a second PI controller is employed for this task, where the input is P_{0ac} and the output is γ, K_{Pp0ac}, and K_{Ip0ac} are needed.

Finally, a deviation from the nominal value (initial condition), m_a, is calculated by a third PI controller. This deviation helps to regulate V_{ac}. The adaptive PI controller input is defined as the difference between the desired and actual voltage magnitude, $V_{ac} - V_{ac}^*$; also, two gain values must be properly specified for the StatCom connected to the electric grid, K_{Pvac} and K_{Ivac}. In total, six gains must be defined for the StatCom controllers. This model captures the main behavior in steady state and dynamic performance of the StatCom. Different to other models presented in the literature for this device, this model includes a phase-shifting transformer and an equivalent shunt susceptance, resulting in an explicit representation of the voltage source converter (VSC) in both sides the AC and DC, respectively. The reader interested in reviewing more details of this model can consult [25].

3. Dynamic Controllers' Gains

In some power systems, a low damping ratio is exhibited. Therefore, the tuning procedure of each controller is a task of precision; moreover, if several gains must be defined, a critical control design stage is presented. An alternative solution for this scenario is to analyze any steady-state condition, and then some gains could be updated online to attain better dynamic performance.

With our strategy it is possible to update all controllers' gains but, to exemplify the relevance of the proposal only some of them are dynamically calculated: the gains for the StatCom K_{Pvac} and K_{Pp0ac} and for each PSS, K_s and T_w. A similar behavior in practice is expected, where only some of them could be retuned by an online procedure. Table 1 exhibits the main steps in proposed control design procedure. The first step consists in use typical gain values obtained around the steady state condition, which are present in Table 2 for the StatCom, and in Table 3 for the generators.

Table 1. Main stages in the proposed procedure to attain adaptive controllers.

Offline and Online Steps in the Proposed Methodology
(i) Offline stage
Extracted key signals from any steady state condition of the power system, input output mapping.
Controllers are initialized to attain this equilibrium point.
An initial architecture of neural networks is defined (shape, size, and learning rate).
An accuracy neural networks architecture is obtained by training procedure, initial dataset.
The proposal is test to different power system scenarios.
(ii) Online stage
The proposed algorithm is updating the controller gains if it required.
The adaptive controllers follow new steady state power system condition.
If disturbances are presented fast changes in controllers gains are exhibited.
Several controllers are tuning simultaneously with positive interaction.

Table 1. *Cont.*

Offline and Online Steps in the Proposed Methodology

(iii) New controllers

If new regulation strategies are included, the adaptive controllers update their self by online learning Equation (11).

Additionally, if it has critic gains, it is possible include new neural networks schemes with a similar architecture, Figure 3.

Table 2. Gains of StatCom controller.

Parameter	Value
$K_{P_E dc}$	1
$K_{I_E dc}$	1.5
$K_{I_p 0 ac}$	0.0015
$K_{I_V ac}$	0.2
R_{ltc}	0 pu
X_{ltc}	0.05
R_{VSC}	2×10^{-3}
X_{VSC}	0.01
G_0	2×10^{-3}

Table 3. Gains of Generator controller.

Generator	K_s	T_w
1	91	10
2	97	9.8
3	100	10
4	100	10

The proposed adaptive PI controller in the StatCom scheme is defined by,

$$u(t) = K_P(e_1)e(t) + K_I(e_1)x_{aux}$$
$$\dot{u}(t) = e(t) \tag{8}$$

For controlling purposes, K_P and K_I and the PSS constants must be defined adequately. We propose to update these gains using the adaptive control law of Figure 3, defined as Equation (9).

$$K_1(e_1) = \beta_1(e_1)\hat{W}_1^T \tag{9}$$

K_1 on the Equation (9), and Figure 3 is used for any of the gains to be calculated.

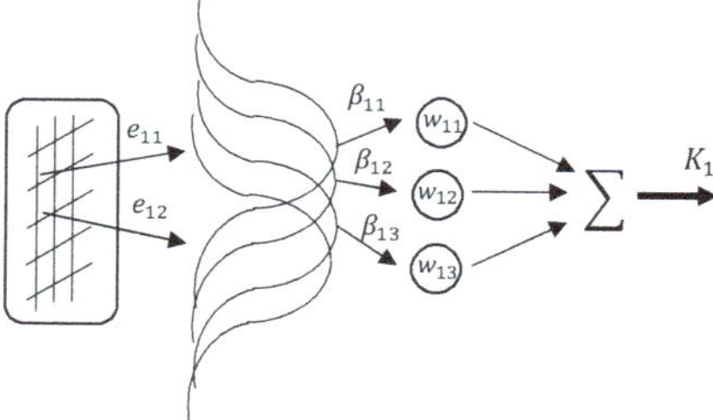

Figure 3. Schematic representation of update procedure.

$|e_1| \leq e^*$ with e^* being a constant. The update law for $\hat{W}_1$ is given by,

$$\hat{W}_1 = W_1 + \Gamma_1(e_1, \beta_1, \varsigma) \tag{10}$$

$\varsigma \in \mathbb{R}$ is a positive constant; $W_1 \in \mathbb{R}^P$ with positive constants and; $\Gamma_1 \in \mathbb{R}^P$ calculated by

$$\Gamma_1 = -\frac{\varsigma \beta_1}{\|\beta_1\|^2} e_1 \tag{11}$$

$\beta_1 \in \mathbb{R}^P$ represents a non-linear relationship from the input signals, e_1. The non-linear relationship is defined by polynomial splines, B-splines [26,27], in this paper are univariate B-spline of fourth order.

Therefore, the controllers update their performance on each sample time with Equations (9)–(11). The input space is normalized in such a way that the input error is bounded in magnitude. First the system is operating offline where learning ratio, ς, is defined in order to get the best performance [28,29]. Then, the dynamic gains are updated by Equation (11) and put to operate online. The results under this last condition are exhibited in Section 4.

The search begins with some typical known values of each gain Equation (12), then the training algorithm is developed to improve the dynamic system response.

$$K_{Pp0ac} = 0.002$$
$$K_{PVac} = 5$$
$$T_1 = T_3 = 0.05$$
$$T_2 = T_4 = 0.01$$
$$V_{min} \leq V_s \leq V_{max}$$
$$V_{max} = 0.05; V_{min} = -0.05 \tag{12}$$

After that, with each operating condition the adaptive algorithm continues learning with input variables and finding the best set of controller gains. The input signals for updating PSS gains are defined by,

$$e_{11} = \omega_i(t) - \omega_B \tag{13}$$
$$e_{12} = P_m - P_e(t) \tag{14}$$

The gains for the StatCom controller have only one input signal, defined by Equation (13). The online procedure consists in calculating the best value for each dynamic gain for the power system operative point. This is possible because the BSNN is updating the weighting vector as a result of input error modification.

Finally, the implementation of the B-Spline neural networks stepwise rules are presented in Table 4, where all mathematical details behind this approach are included.

In this work, B-Spline neural networks algorithm was selected because it requires less computational effort, thanks to its single layer of neurons, its structure, and the shape of the base functions, Figure 3, in contrast to the multi-layer neural networks architecture. Furthermore, the activation functions are linear with respect to the adaptive weights, with an instant learning rule that can be used to update and adjust the weights online. These conditions make the B-Spline neural networks algorithm able of modeling and regulating complex non-linear systems. With these features, a robust, optimal control system is obtained with the ability to be adapted to inherent non-linearities and external or internal disturbances of the system. One of the core aspects of selecting the use of the BSNN is that by defining the base functions a non-linear relation of the input is obtained, and the training algorithm is computationally efficient, with a numerically stable recurring relationship that works with any distribution of knot vector.

Table 4. 1: B-Spline Neural Network off-line training rules.

Input Define:	space lattice with n knot-vectors
Define:	basis functions (K order, shape and distribution)
Define:	number of knot-vector
Define:	nodes of hidden layer
	Define: Initial conditions (weights)
	Define: error signal and minimum and maximum values
	Define: threshold error
	while t < simulation time do
	Calculate the input and output value of each layer
	Calculate the errors between target and current value
	Includes several operating conditions
	if e_x < threshold error
	NN is ready for online operation
	return weights
	else
	Update weights
	Calculate the input and output value of each layer
	Calculate the errors between target and actual value
	if e_x < threshold error
	NN is ready for online operation
	return weights
	else
	Update weights
	Change data source and update learning rate
	end if
	initialize the process
	end if
	return weights, K order, threshold error
	end
	on-line training
	Load weights, K order, threshold error, error signal
	while t < simulation time do
	Calculate the input and output value of each layer
	Calculate the errors between target and current value
	if e_x < threshold error
	return k_1
	else
	Update weights ec. Equation (10)
	Calculate the input and output value of each layer, ec.
	Equation (9)
	return k_1
	end

4. Test Power System

Without loss of generality, the performance of our proposal is proved in an important benchmark of four machine and two area electric power system [1]. Although this system is not big, it presents an interesting and complex behavior for transient stability studies. The two areas are connected by a weak tie, this is an important case for studying the fundamental nature of inter-area oscillations and the inherent difficulty in tuning and controllers coordination.

4.1. Case Base

A base analysis for different types of excitation control is presented in [1]. This power system has three rotor angle modes of oscillation. There is one inter-area mode of 0.55 Hz with generator 1 and 2 swinging against generators 3 and 4 of area 2. Two more are intermachine oscillation local modes, one of 1.09 Hz corresponds to area 1 and the second of 1.12 Hz is for area 2. These modes are determined only with one type of excitation control that changes if the operating condition varies or more dynamic components are involved.

In this analysis, each synchronous generator of Figure 4 is represented by the model Equation (1). Two loads are connected at bus 7 and 9. The initial conditions are based in data set reported in [1]. The active power injected by each generator: $P_{G1} = P_{G2} = P_{G4} = 700$ MW and, $P_{G3} = 719$ MW; terminal voltages: $V_{tG1} = 1.03\angle 20.2$ pu, $V_{tG2} = 1.01\angle 10.5$ pu, $V_{tG3} = 1.03\angle - 6.8$ pu, $V_{tG4} = 1.01\angle - 17.0$ pu. The initial conditions for state variables of each generator are obtained considering (1)–(4) using this information. The machine parameters are also available in [1] and used on this paper.

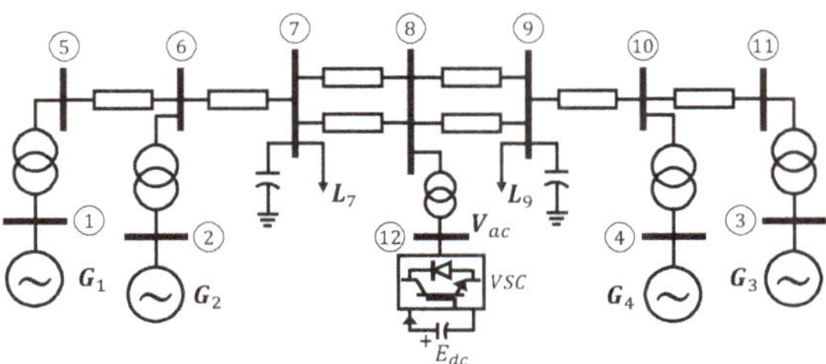

Figure 4. Four machine power system with StatCom at bus 8.

With this model representation and initial conditions, the system is exposed to a three-phase fault at different nodes with similar results, selecting node 7 on this section to illustrate the results.

The electrical grid without StatCom neither PSS exhibits an unstable performance when the fault is cleared after eight cycles, Figure 5. If the fault is cleared up to seven cycles a stable evolution is observed, but the oscillations have values far from the prefault condition, and with long duration. After 6 s the oscillations continue with very little damping. Figure 5 shows the angular difference between machines with generator number one as a reference, for eigth cycles (unstable operation) and six cycles (stable but oscillating operation). Additional damping required is evidenced. Similar behavior is observed in other system variables.

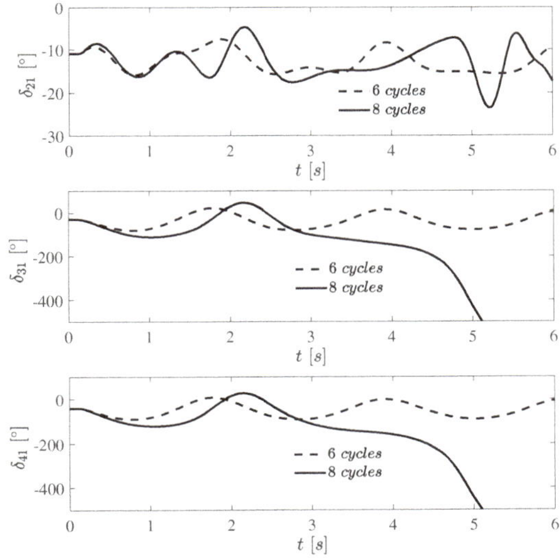

Figure 5. Angular difference of each machine with respect to number one, electrical grid without StatCom neither PSS.

4.2. Case 1. Three-Phase Failure Cleared after Eight Cycles

This study considers a StatCom integrated to the power system at bus 8, as presented in Figure 4. Initially, the StatCom operation was forced to keep the system behavior very similar to one without StatCom and PSS. Therefore, the power flow solution is used to feed the calculation of initial conditions by the prefault situation of angles and voltage magnitude. At bus 8, the voltage magnitude is $V_8 = 0.9556$ pu, close to the lower limit.

A three phase fault at 0.1 s is presented near to node 7 in one of the transmission lines 7–8. Several fault duration times were tested. Figures 6 and 7 exhibit the power system performance when the StatCom and PSS have a positive interaction. The fault is cleared after eight cycles, which was the fault duration for the power system to become unstable on the base case. There is a comparison using StatCom with and without PSS for making the system stable. The StatCom inclusion is not enough to improve the global power system damping, however, it diminishes the magnitude of the oscillations and improves the damping ratio respect to the system without controllers in the same period.

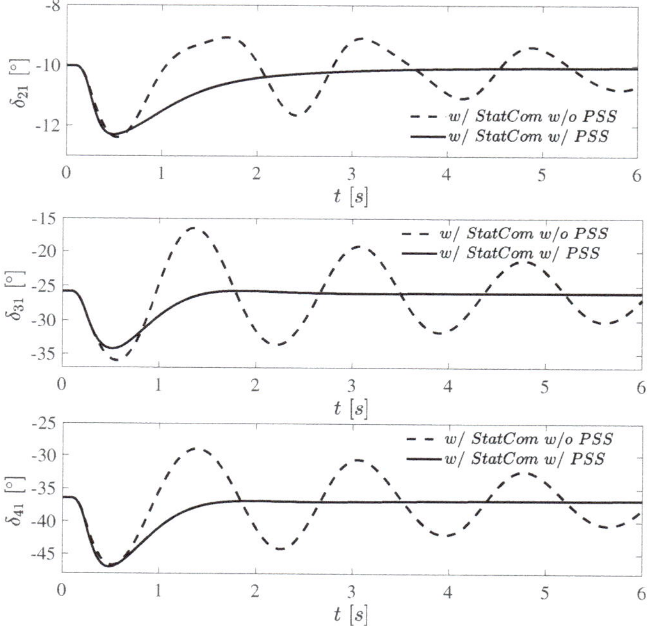

Figure 6. Angular difference of each machine with respect to number one, case 1: 8 cycles until the fault is cleared.

Figure 6 depicts the angular difference of each generator. The proposed control design methodology shows a positive interaction between controllers and the overshoot has an important reduction with and without the PSS, but the oscillations are eliminated in a fast way when PSS is included. Figure 7a presents the active power on the generator two and Figure 7b the voltage magnitude in PCC node where the StatCom is included.

Table 5 presents the main characteristics of transient response for case 1. Where the quantitative comparison is obtained in the case: (i) with StatCom and without PSS (wS/woPSS); and (ii) with StatCom and with PSS (wS/wPSS). The transient responses of the angular differences are improved with the correct coordination of controllers. In the case of δ_{21} the settling time is diminished in about 80.5%, for the overshot a marginal improves percent is obtained. However, the overall performance of the proposed technique permits to attain a similar behavior in terms of overshot, in some case better, but in all responses the settling time is drastically enhanced.

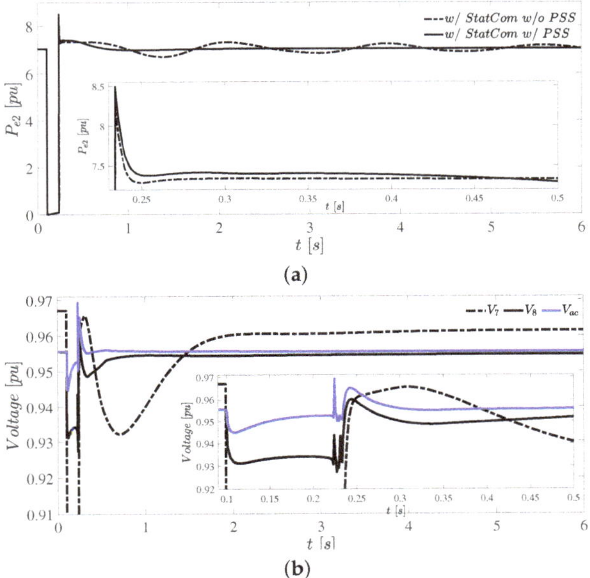

Figure 7. Case 1: (a) real power at generator 2; (b) voltage magnitude in bus of StatCom connection.

Table 5. Analysis in time domain of the responses for case 1.

Variable	Case	Rise Time	Settling Time	Overshoot	Peak Time
δ_{21}	wS/woPSS	0.0802	14.4562	2.3927	0.5300
	wS/wPSS	0.0198	3.8236	2.2998	0.4960
δ_{31}	wS/woPSS	0.0515	16.1978	10.1564	0.5670
	wS/wPSS	0.0205	2.2131	8.5395	0.5180
δ_{41}	wS/woPSS	0.0571	15.7923	10.2529	0.5120
	wS/wPSS	0.0238	2.8591	10.3163	0.4820
P_{e2}	wS/woPSS	0.00054	4.0704	1.2251	0.2340
	wS/wPSS	0.00052	0.5917	1.4949	0.2340

The settling time is improved with the following percentages; for δ_{31}, 86.3%; for δ_{41}, 81.9%, and P_{e2}, 85.5%, which is concentrated in Table 6. The rise time and peak time are similar for both controller's tuning. Like the previous dynamic performance of the power system, for the case 2, also the transient response features are determined, Table 6. Now, the proposed adaptive strategy has impacted in two main features in time domain transient response. Both settling time and overshoot are clearly enhanced by the proposed control coordination scheme. The following improvement values are attained: δ_{21}, 83%; δ_{31}, 87.5%; δ_{41}, 86%, and P_{e2} in 85%. In Table 6, the results are presented. It is evident the correct performance of the proposed algorithm to diminish the exhibited low frequency oscillations in a faster way. Additionally, the percent that diminishes the overshoot in this case is: δ_{21}, 65%; δ_{31}, 70%; δ_{41}, 79%, and P_{e2} in 43.3%.

Table 6. Analysis in time domain of the responses for case 2.

Variable	Case	Rise Time	Settling Time	Overshoot	Peak Time
δ_{21}	wS/woPSS	0.1568	9.7201	2.1921	1.2450
	wS/wPSS	0.1339	1.6374	0.4907	0.3415
δ_{31}	wS/woPSS	0.3151	17.6893	12.6784	1.0090
	wS/wPSS	0.4666	2.2153	1.5931	1.1230
δ_{41}	wS/woPSS	0.3149	14.7150	10.9615	1.0640
	wS/wPSS	0.3860	2.0571	2.2949	1.0620
P_{e2}	wS/woPSS	0.0026	4.9918	0.4435	0.8310
	wS/wPSS	0.0019	0.7363	0.2515	0.7930

4.3. Case 2. Three-Phase Failure with Line Out of Service after Eight Cycles

In contrast to case 1, now the postfault condition is with one of the parallel transmission lines 7–8 out of service after the fault is cleared. The fault with six cycles of duration is presented between buses 7–8, close to node 7. The proposed methodology has a behavior similar to case 1, Figures 8 and 9.

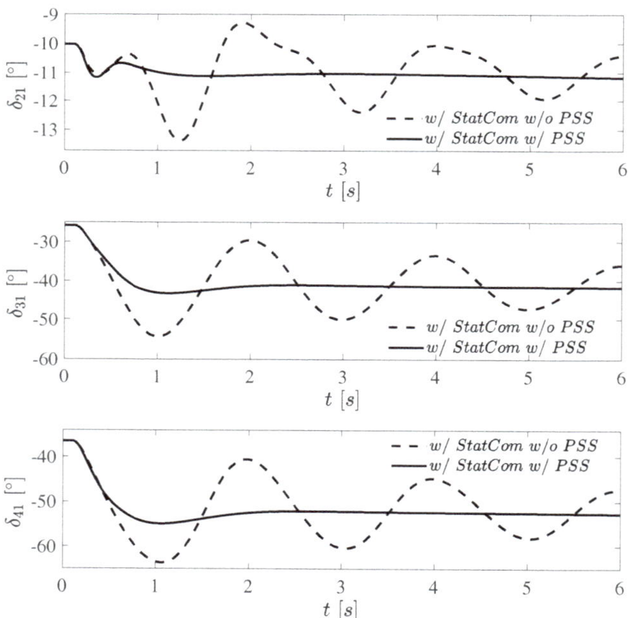

Figure 8. Angular difference of each machine with respect to number one, case 2.

The angular difference of machines respect to number one are presented in Figure 8. The overshot after the fault is released is bounded and the oscillations of internal machine angle exhibit fast transient respond reaching a new value in steady state. The final condition is due to a new electrical grid topology with one transmission line out of service.

Figure 9a reveals the active power injected by the generator 2. With improved control design stage and dynamic gains, the controllers are adapted to the new power system condition or perturbation presented. The power oscillations are diminished after one second. With only StatCom the oscillations are diminished with low magnitude.

In Figure 9b some voltage magnitudes are presented. At node 8 where StatCom is connected the voltage magnitude returns to the set point, however, it is close to low voltage limit. In the case of faulted node 7, the system tries to return a stable condition, but the voltage is below low limit. Thus, an action of secondary control loop is required to reach the new steady state condition where all variables should be within physical limits.

The evolution of the gains is exhibited in Figure 10a, the updated values allow to get the best performance. All results are in accordance with the expected values of the power system with improved damping ratio due to the design procedure and the inclusion of some dynamic gains. Additionally, the performance of these gains in case 2 is presented in Figure 10a, the initial conditions are equal for both study cases but have different dynamic evolution.

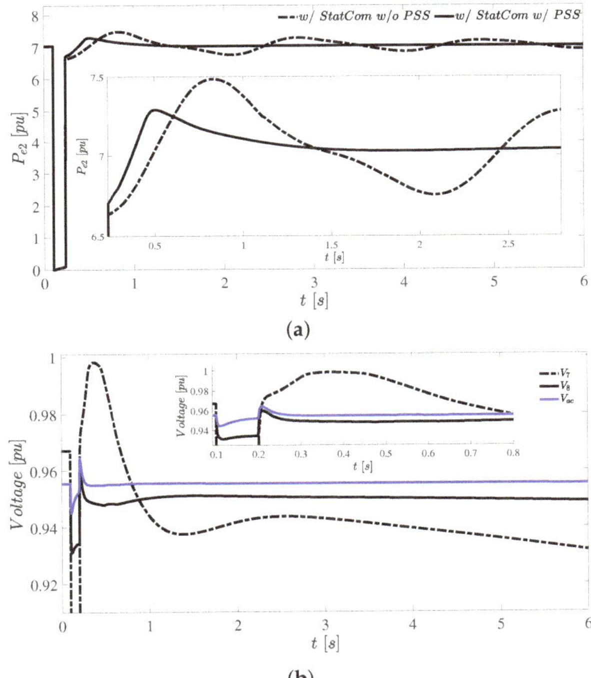

(a)

(b)

Figure 9. Case 2: (**a**) real power at generator 2; (**b**) voltage magnitude in bus of StatCom connection.

In Figure 10b, the control signals for the StatCom are displayed. The calculated gains are meant to get a fast response with limited overshoot, after the transient period both control signals attain a new steady state condition. Similar behavior for both cases is exhibited in these signals.

The deviation respect to initial values is small, and with very fast response (less than one second).

Cases 1 and 2 demonstrate the improved dynamic performance that the power system exhibits by using the variables of this section. The simulation results indicate that critical clearing time has been improved widely. Under this scenario the presented results illustrate the system response when the fault is cleared after six cycles. The prefault and postfault condition are the same.

Then, without loss of generality it is possible to extend our methodology in electric power system with more generators, different FACTS devices, generation plants and emerging technologies. Under these new conditions some minor considerations must be included in B-Spline scheme for on line operation.

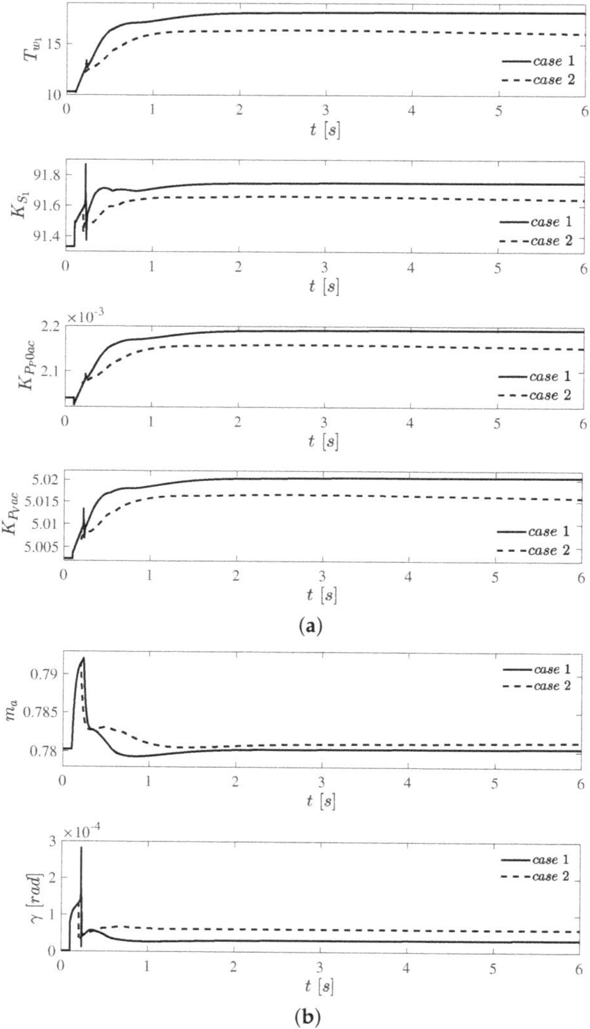

Figure 10. (**a**) Evolution of dynamic gains and; (**b**) Control signals for the StatCom.

5. Conclusions

The proposed design methodology includes a complete non-linear representation of large-scale power systems for transient stability studies. Three-phase failure is presented to validate stability. On the base case, the angular difference of the machines is not ensured when failure is released after 8 cycles. The power system presents enhanced dynamic performance when StatCom and PSS are included. The PSS and the StatCom controllers are simultaneously tuned using B-Spline for gain definition, and using typical known values for each gain. Model uncertainties are not included in the design stage because they are considered on the gains updating task, which is performed in each sample time, avoiding model and parameter dependency. The control design stage allows good system performance under a specific operating point but also with other operating conditions or topologies. Moreover, the new proposed technique can be extended to other complex multimachine power systems with several adaptive dynamic controllers.

Author Contributions: Conceptualization, R.T.-O.; Data curation, O.A.-M.; Formal analysis, F.B.-C.; Investigation, R.T.-O.; Methodology, R.T.-O., F.B.-C. and A.V.-G.; Supervision, F.B.-C. and A.V.-G.; Validation, O.A.-M.; Writing–original draft, R.T.-O.; Writing–review–editing, A.V.-G. All authors have read and agreed to the published version of the manuscript.

Funding: This research received no external funding.

Institutional Review Board Statement: Not applicable.

Informed Consent Statement: Not applicable.

Data Availability Statement: No new data were created or analyzed in this study. Data sharing is not applicable to this article.

Acknowledgments: R.T.-O. and O.A.-M. thank to the CONACYT-SENER-SUSTENTABILIDAD ENERGÉTICA the support to develop this work under grant CEMIE-REDES PE-A-21. A.V.-G. thanks Universidad Panamericana the support to develop this work. R.T.-O. thanks to the Programa de Apoyo a Proyectos de Investigación e Innovación Tecnológica (PAPIIT) de la UNAM. Con clave UNAM-DGAPA-PAPIIT- IA106920.

Conflicts of Interest: The authors declare no conflict of interest.

References

1. Kundur, P.; Balu, N.; Lauby, M. *Power System Stability and Control*; McGraw-Hill Education: New York, NY USA, 1994; Volume 7.
2. Kang, R.D.; Martinez, E.A.; Viveros, E.C. Coordinated tuning of power system controllers using parallel genetic algorithms. *Electr. Power Syst. Res.* **2021**, *190*, 106628. [CrossRef]
3. Liu, M.; Bizzarri, F.; Brambilla, A.M.; Milano, F. On the Impact of the Dead-Band of Power System Stabilizers and Frequency Regulation on Power System Stability. *IEEE Trans. Power Syst.* **2019**, *34*, 3977–3979. [CrossRef]
4. Rana, M.J.; Shahriar, M.S.; Shafiullah, M. Levenberg–Marquardt neural network to estimate UPFC-coordinated PSS parameters to enhance power system stability. *Neural Comput. Appl.* **2019**, *31*, 1237–1248. [CrossRef]
5. Pandey, R.K.; Gupta, D.K. Integrated multi-stage LQR power oscillation damping FACTS controller. *CSEE J. Power Energy Syst.* **2018**, *4*, 83–91. [CrossRef]
6. Bayatloo, F.; Bozorgi-Amiri, A. A novel optimization model for dynamic power grid design and expansion planning considering renewable resources. *J. Clean. Prod.* **2019**, *229*, 1319–1334. [CrossRef]
7. Xu, T.; Birchfield, A.B.; Overbye, T.J. Modeling, tuning, and validating system dynamics in synthetic electric grids. *IEEE Trans. Power Syst.* **2018**, *33*, 6501–6509. [CrossRef]
8. Rimorov, D.; Kamwa, I.; Joós, G. Model-based tuning approach for multi-band power system stabilisers PSS4B using an improved modal performance index. *IET Gener. Transm. Distrib.* **2015**, *9*, 2135–2143. [CrossRef]
9. Talaq, J. Optimal power system stabilizers for multi machine systems. *Int. J. Electr. Power Energy Syst.* **2012**, *43*, 793–803. [CrossRef]
10. Salgotra, A.; Pan, S. Model based PI power system stabilizer design for damping low frequency oscillations in power systems. *ISA Trans.* **2018**, *76*, 110–121. [CrossRef]
11. Gomes, S., Jr.; Guimarães, C.; Martins, N.; Taranto, G. Damped Nyquist Plot for a pole placement design of power system stabilizers. *Electr. Power Syst. Res.* **2018**, *158*, 158–169. [CrossRef]
12. Liu, X.; Han, Y. Decentralized multi-machine power system excitation control using continuous higher-order sliding mode technique. *Int. J. Electr. Power Energy Syst.* **2016**, *82*, 76–86. [CrossRef]
13. Patel, A.; Ghosh, S.; Folly, K.A. Inter-area oscillation damping with non-synchronised wide-area power system stabiliser. *IET Gener. Transm. Distrib.* **2018**, *12*, 3070–3078. [CrossRef]
14. Sands, T. Development of Deterministic Artificial Intelligence for Unmanned Underwater Vehicles (UUV). *J. Mar. Sci. Eng.* **2020**, *8*, 578. [CrossRef]
15. Cooper, M.; Heidlauf, P. Nonlinear feed forward control of a perturbed satellite using extended least squares adaptation and a luenberger observer. *J. Aeronaut. Aerosp. Eng.* **2018**, *7*, 1–7. [CrossRef]
16. Shi, Y.; Sarlioglu, B.; Lorenz, R.D. Real-time loss minimizing control of induction machines for dynamic load profiles under deadbeat-direct torque and flux control. *IEEE Trans. Ind. Appl.* **2021**, *1*, 1–9.

17. Yao, W.; Jiang, L.; Wen, J.; Wu, Q.; Cheng, S. Wide-area damping controller for power system interarea oscillations: A networked predictive control approach. *IEEE Trans. Control Syst. Technol.* **2014**, *23*, 27–36. [CrossRef]
18. Farahani, M.; Ganjefar, S. Intelligent power system stabilizer design using adaptive fuzzy sliding mode controller. *Neurocomputing* **2017**, *226*, 135–144. [CrossRef]
19. Yousefian, R.; Kamalasadan, S. A Lyapunov function based optimal hybrid power system controller for improved transient stability. *Electr. Power Syst. Res.* **2016**, *137*, 6–15. [CrossRef]
20. Ray, P.K.; Paital, S.R.; Mohanty, A.; Eddy, F.Y.; Gooi, H.B. A robust power system stabilizer for enhancement of stability in power system using adaptive fuzzy sliding mode control. *Appl. Soft Comput.* **2018**, *73*, 471–481. [CrossRef]
21. Afzalan, E.; Joorabian, M. Analysis of the simultaneous coordinated design of STATCOM-based damping stabilizers and PSS in a multi-machine power system using the seeker optimization algorithm. *Int. J. Electr. Power Energy Syst.* **2013**, *53*, 1003–1017. [CrossRef]
22. Wang, S.K. Coordinated parameter design of power system stabilizers and static synchronous compensator using gradual hybrid differential evaluation. *Int. J. Electr. Power Energy Syst.* **2016**, *81*, 165–174. [CrossRef]
23. Al-Ismail, F.; Hassan, M.; Abido, M. RTDS implementation of STATCOM-based power system stabilizers. *Can. J. Electr. Comput. Eng.* **2014**, *37*, 48–56. [CrossRef]
24. Padiyar, K. *Power System Dynamics: Stability and Control*; Anshan: Kent, UK, 2004.
25. Castro, L.M.; Acha, E.; Fuerte-Esquivel, C.R. A novel STATCOM model for dynamic power system simulations. *IEEE Trans. Power Syst.* **2013**, *28*, 3145–3154. [CrossRef]
26. Brown, M.; Harris, C.J. *Adaptive B-Spline Networks, Neurofuzzy Adaptive Modelling and Control*; Prentice Hall: New York, NY, USA, 1994.
27. Toshniwal, D.; Speleers, H.; Hiemstra, R.R.; Manni, C.; Hughes, T.J. Multi-degree B-splines: Algorithmic computation and properties. *Comput. Aided Geom. Des.* **2020**, *76*, 101792. [CrossRef]
28. Tapia, R.; Aguilar, O.; Minor, H.; Santiago, C. Power system stabilizer and secondary voltage regulator tuning for multi-machine power systems. *Electr. Power Components Syst.* **2012**, *40*, 1751–1767. [CrossRef]
29. Beltran-Carbajal, F.; Tapia-Olvera, R.; Lopez-Garcia, I.; Guillen, D. Adaptive dynamical tracking control under uncertainty of shunt DC motors. *Electr. Power Syst. Res.* **2018**, *164*, 70–78. [CrossRef]

Article

A Novel Exact Plate Theory for Bending Vibrations Based on the Partial Differential Operator Theory

Chuanping Zhou [1,2,3], Maofa Wang [1,*], Xiao Han [3], Huanhuan Xue [1], Jing Ni [1] and Weihua Zhou [4]

[1] School of Mechanical Engineering, Hangzhou Dianzi University, Hangzhou 310018, China; zhoucp@hdu.edu.cn (C.Z.); weiyangchaohu@163.com (H.X.); nj2000@hdu.edu.cn (J.N.)
[2] School of Mechatronics Engineering, University of Electronic Science and Technology of China, Chengdu 611731, China
[3] Hangzhou Changchuan Technology Co., Ltd., Hangzhou 310018, China; hanxiao@hzcctech.net
[4] College of Electrical Engineering, Zhejiang University, Hangzhou 310027, China; dididi@zju.edu.cn
* Correspondence: wmf@hdu.edu.cn; Tel.: +86-134-2916-8099

Abstract: Thick wall structures are usually applied at a highly reduced frequency. It is crucial to study the refined dynamic modeling of a thick plate, as it is directly related to the dynamic mechanical characteristics of an engineering structure or device, elastic wave scattering and dynamic stress concentration, and motion stability and dynamic control of a distributed parameter system. In this paper, based on the partial differential operator theory, an exact elasto-dynamics theory without assumptions for bending vibrations is presented by using the formal solution proposed by Boussinesq–Galerkin, and its dynamic equations are obtained under appropriate gauge conditions. The exact plate theory is then compared with other theories of plates. Since the derivation of the dynamic equation is conducted without any prior assumption, the proposed dynamic equation of plates is more exact and can be applied to a wider frequency range and greater thickness.

Keywords: exact plate theory; thick plate; bending vibration; partial differential operator theory; gauge condition

Citation: Zhou, C.; Wang, M.; Han, X.; Xue, H.; Ni, J.; Zhou, W. A Novel Exact Plate Theory for Bending Vibrations Based on the Partial Differential Operator Theory. *Mathematics* **2021**, *9*, 1920. https://doi.org/10.3390/math9161920

Academic Editor: Francisco Beltran-Carbajal

Received: 24 May 2021
Accepted: 2 August 2021
Published: 12 August 2021

Publisher's Note: MDPI stays neutral with regard to jurisdictional claims in published maps and institutional affiliations.

1. Introduction

As typical structures, thick plates with holes are widely used under dynamic loads in aerospace, ocean engineering, civil engineering and mechanical engineering [1–5]. Due to the stress concentration near the load-bearing opening in the finite thickness structure, there will be an intense three-dimensional effect zone. The three-dimensional effect zone is closely related to the relative thickness of the structure, which largely controls the fracture, fatigue and other mechanical properties of the structure [6,7]. There is a large error in calculating the dynamic problem of the plate with actual thickness based on the classical thin plate theory. With the development of modern science and technology, engineering structure designs tend to be light, and the way to achieve this is to use advanced materials and improve the structure design theory. However, because the three-dimensional problem has not been well solved, it will encounter great mathematical difficulties in solving the three-dimensional problem.

The classical plate theory (CPT) is the basic theory in the hierarchy of plate theories [6]. Since it was formulated systematically in the 19th century, CPT has been widely applied for buckling, bending and vibration analyses of plates. However, it is worth noting that CPT does not consider the effects of shear deformation or rotary inertia [8–10]. Hence, in the analysis of thick plates and also in the case of thin plates vibrating at higher frequencies, the use of CPT would result in considerable errors [11–14]. These limitations of CPT led to the development of first-order and higher-order shear deformation plate theories.

Reissner [11,12] proposed a thick plate theory, including the effects of shear deformation. Reissner's theory involves three coupled governing differential equations in terms of three unknowns. In fact, Reissner's theory is based on a stress approach. Mindlin [13] considered both the effects of shear deformation and rotary inertia and proposed a first-order displacement-based theory, which involves three coupled governing differential equations. Mindlin's theory was derived by using the frequency domain method, so its disadvantage is that it cannot predict the constant transverse shear strains and stresses across the plate thickness. Therefore, the theory requires a shear correction factor to match the strain energy calculated by using constant transverse shear strains and stresses. The transverse shear strains and stresses vary parabolically across the plate thickness [14].

The Levinson [15] and Reddy [16,17] plate theories are higher-order shear deformation plate theories based on displacement. Both the theories are based on the same displacement field. However, the governing equations of the Levinson theory are derived from the plate gross equilibrium equations, whereas Reddy used the principle of virtual displacements to derive the governing differential equations. Neither theory requires a shear correction factor. The plate theory proposed by Kant [18] is based on a higher-order displacement model, which causes a secondary change in transverse shear strain and a linear change in transverse normal strain across the plate thickness. This work involves the flexure of thick rectangular isotropic plates. The formulation of the theory involves six variables. Since the higher-order plate theories cannot take the continuity conditions of displacement and shear stresses into consideration, the accuracy of analytical solutions cannot meet the requirements in engineering applications [19].

Nedri [20] presented a novel refined hyperbolic shear deformation theory based on the assumption that the transverse displacements consist of shear and bending components where the bending components do not contribute to shear forces, and likewise, the shear components do not contribute to bending moments. The irrelevancies of the two components make calculations simple, yet it may lead to an erroneous result when the displacements vary sharply across the thickness. Shimpi [21] developed a refined plate theory (RPT) with only two unknown functions available in the paper. RPT produces two fourth-order governing differential equations, which are uncoupled for static problems and are only inertially coupled for dynamic problems. Subsequently, Shimpi [22] presented two new first-order shear deformation plate theories with only two unknown functions to improve RPT. Using the method proposed by Timoshenko and Ashwell, Nicassio [4] presented a novel forecast model to map the surface profiles of bistable laminates and developed an analytical model to provide an interpretation of the bistable shapes in terms of principal and anticlastic curvatures. Wu [5] gave a revised method to increase the stiffness and natural frequency of bistable composite shells, which can be suitable for spatial, lightweight structural components.

Among various refined plate theories, Carrera [23] presented a unified expression, which provides a program to obtain refined structural models of beams, plates and shells that explain variable kinematics descriptions. Based on Carrera's unified formulation (CUF), these structural models are obtained using the N-order Taylor expansion to expand the unknown displacement variables. Tornabene et al. [24,25] derived a general formulation of 2D higher-order equivalent plate theory. The theoretical framework covers the static and dynamic analysis of shell structures by using a general displacement field based on CUF. In the CUF system, a linear case can describe a classical model, while a higher-order case can describe a three-dimensional structure. Kolahchi [3] investigated bending, buckling and buckling of embedded nano-sandwich plates based on refined zigzag theory (RZT), sinusoidal shear deformation theory (SSDT), first-order shear deformation theory (FSDT) and classical plate theory (CPT), and a differential cubature (DC) method is applied for obtaining the static response, the natural frequencies and the buckling loads of nano-sandwich plates. The numerical investigation shows that RZT is highly accurate in predicting the deflection, frequency and buckling load of nano-sandwich plates without requiring any shear correction factors.

In brief, although the above CPT theories have been optimized and updated, they are theoretically based on the geometric method, and the models are rough since engineering assumptions are still used during the derivation, which results in many limitations in the application of thick wall structures, especially in the case of plates vibrating at higher frequencies. In this paper, we propose a novel theory of exact elastic dynamics for bending plates not based on the geometric view but the algebraic view. During the derivation, we apply the general formal solution proposed by Boussinesq–Galerkin and the operator theory of partial differential equations. The exact elasto-dynamics equations for bending plates are obtained by using appropriate gauge conditions, and the exact dynamic theory of thick plates is compared with other plate theories. Since the derivation of the dynamic equation is carried out without any prior assumption, the proposed dynamic equation of plates is more exact and can be applied in a wider frequency range and greater thickness. The exact thick plate theory in this paper makes up for the shortcomings of the classical thin plate theory and other thick plate theories. It can be used not only for structures with large thickness span ratio but also for vibration mechanics problems with great influence of shear deformation and moment of inertia, such as spacecraft attitude dynamics and control, structural motion stability, dynamic stress concentration for thick plates with holes, and calculations for submarine anechoic tile structure design.

2. Derivation of the Exact Dynamic Theory for the Bending Plate

First, the derivation process of the exact plate theory for the bending plate is introduced. According to the three-dimensional elasto-dynamics theory, the governing equation of the spatial displacement field is the Navier equation:

$$\mu \nabla^2 \mathbf{u} + (\lambda + \mu) \nabla (\nabla \cdot \mathbf{u}) = \rho \frac{\partial^2 \mathbf{u}}{\partial t^2}, \tag{1}$$

where μ, λ are the Lame constants, $\nabla = \mathbf{i}\partial/\partial x + \mathbf{j}\partial/\partial y + \mathbf{k}\partial/\partial z$, ρ is the density.

From Equation (1), based on Boussinesq–Galerkin solution (B-G solution), the solution given as:

$$\mathbf{u} = 2(1 - \nu) \left(\nabla^2 - \frac{1}{c_1^2} \frac{\partial^2}{\partial t^2} \right) \mathbf{G} - \nabla (\nabla \cdot \mathbf{G}), \tag{2}$$

where c_1, c_2 are longitudinal wave velocity and transverse wave velocity, ν is the Poisson ratio, and $\mathbf{G} = (G_1, G_2, G_3)$ is the Somigliana vector potential function, which satisfies the following relation as:

$$\prod \left(\nabla^2 - T_j^2 \right) \mathbf{G} = 0. \tag{3}$$

where T_j are time differential operators, $T_j^2 = \frac{1}{c_j^2} \frac{\partial^2}{\partial t^2}, (j = 1, 2)$,

Using the Taylor series expansion of the exponential operator function, the displacement at any point in the plate can be written as:

$$u_x(x, y, z) = \exp \left(z \frac{\partial}{\partial z} \right) u_x(x, y, 0), \tag{4}$$

$$u_y(x, y, z) = \exp \left(z \frac{\partial}{\partial z} \right) u_y(x, y, 0), \tag{5}$$

$$u_z(x, y, z) = \exp \left(z \frac{\partial}{\partial z} \right) u_z(x, y, 0). \tag{6}$$

The fluctuation of plate bending is a case of antisymmetric motion, and Equation (4) can be written as:

$$u_x(x, y, z) = \sin h \left(z \frac{\partial}{\partial z} \right) u_x(x, y, 0), \tag{7}$$

$$u_y(x, y, z) = \sin h \left(z \frac{\partial}{\partial z} \right) u_y(x, y, 0), \tag{8}$$

$$u_z(x,y,z) = \cosh\left(z\frac{\partial}{\partial z}\right)u_z(x,y,0),\tag{9}$$

where $\sinh(\cdot)$ is hyperbolic sine function, and $\cosh(\cdot)$ is a hyperbolic cosine function.

The B-G solution can be written as:

$$G_j(x,y,z)= \sinh\left(z\frac{\partial}{\partial z}\right)G_j(x,y,0) = \cosh\left(z\frac{\partial}{\partial z}\right)\sum_{i=1}^{2}G_j^i(x,y,0),\tag{10}$$

$$G_3(x,y,z)= \cosh\left(z\frac{\partial}{\partial z}\right)G_3(x,y,0) = \cosh\left(z\frac{\partial}{\partial z}\right)\sum_{i=1}^{2}G_3^i(x,y,0),\tag{11}$$

The Somigliana vector potential function $\mathbf{G}$ can be decompose into two vector potential function as $\mathbf{G} = \mathbf{G}^1 + \mathbf{G}^2$, and $\left(\nabla_j^2 + \frac{\partial^2}{\partial z^2}\right)\mathbf{G}^j = 0$, here $\nabla_j^2 = \nabla^2 - T_j^2, (j=1,2)$ is the Lorentz operator. The trigonometric function operator can be written as:

$$\frac{\sin(z\nabla_j)}{\nabla_j} = \sum_{n=0}^{\infty}(-1)^n\frac{1}{(2n+1)!}z^{2n+1}\nabla_j^{2n},\tag{12}$$

$$\cos(z\nabla_j) = \sum_{n=0}^{\infty}(-1)^n\frac{1}{(2n)!}z^{2n}\nabla_j^{2n},\tag{13}$$

here $j = 1,2$.

We can also obtain the following relation as:

$$\nabla_0\mathbf{G} = \frac{\partial G_1}{\partial x} + \frac{\partial G_2}{\partial y} + \frac{\partial G_3}{\partial z} = \sum_{j=1}^{2}\frac{\sin(z\nabla_j)}{\nabla_j}\left(\frac{\partial g_1^j}{\partial x} + \frac{\partial g_2^j}{\partial y} - \nabla_j^2 g_3^j\right).\tag{14}$$

For the sake of avoiding the non-uniqueness of unknown functions, two gauge conditions are adopted as follows:

$$\frac{\partial g_1^j}{\partial x} + \frac{\partial g_2^j}{\partial y} = 0, (j=1,2).\tag{15}$$

Equation (8) can be written as:

$$\begin{aligned}
\nabla_0(\nabla_0\cdot\mathbf{G}) =&-\sum_{j=1}^{2}\left[\frac{\sin(z\nabla_j)}{\nabla_j}\nabla_j^2\frac{\partial}{\partial x}g_3^j\right]\mathbf{i} - \sum_{j=1}^{2}\left[\frac{\sin(z\nabla_j)}{\nabla_j}\nabla_j^2\frac{\partial}{\partial y}g_3^j\right]\mathbf{j}\\
&-\sum_{j=1}^{2}\left[\cos(z\nabla_j)\nabla_j^2 g_3^j\right]\mathbf{k}
\end{aligned}\tag{16}$$

The displacement in the plate can be expressed as:

$$\mathbf{u} = 2(1-\nu)\left(\nabla_1^2 + \frac{\partial^2}{\partial z^2}\right)\mathbf{G} - \nabla_0(\nabla_0\cdot\mathbf{G}).\tag{17}$$

Its component-wise expressions can be written as:

$$u_x = \frac{\sin(z\nabla_2)}{\nabla_2}T_2^2 g_1^2 + \sum_{j=1}^{2}\frac{\sin(z\nabla_j)}{\nabla_j}\nabla_j^2\frac{\partial}{\partial x}g_3^j,\tag{18}$$

$$u_y = \frac{\sin(z\nabla_2)}{\nabla_2}T_2^2 g_2^2 + \sum_{j=1}^{2}\frac{\sin(z\nabla_j)}{\nabla_j}\nabla_j^2\frac{\partial}{\partial y}g_3^j,\tag{19}$$

$$u_z = \cos(z\nabla_2)T_2^2 g_3^2 + \sum_{j=1}^{2} \cos(z\nabla_j)\nabla_j^2 g_3^j. \tag{20}$$

Considering the neutral surface displacement and the normal angle, the generalized displacement in the plate can be expressed as:

$$\psi_x = -\left.\frac{\partial u_x}{\partial z}\right|_{z=0} = -T_2^2 g_1^2 - \sum_{j=1}^{2} \nabla_j^2 \frac{\partial^2 g_3^j}{\partial x}, \tag{21}$$

$$\psi_y = -\left.\frac{\partial u_y}{\partial z}\right|_{z=0} = -T_2^2 g_2^2 - \sum_{j=1}^{2} \nabla_j^2 \frac{\partial^2 g_3^j}{\partial y}, \tag{22}$$

$$w = \left. u_z\right|_{z=0} = \nabla^2 g_3^2 + \nabla_1^2 g_3^1. \tag{23}$$

The rotational normal angle to the neutral surface can be expressed as:

$$\psi_x = \frac{\partial F}{\partial x} + \frac{\partial f}{\partial y}, \quad \psi_y = \frac{\partial F}{\partial y} - \frac{\partial f}{\partial x}. \tag{24}$$

The functions g_1^2, g_2^2, g_3^2 can be expressed by the neutral surface displacement and normal angle as:

$$g_1^2 = -T_2^{-2}\left(\frac{\partial f}{\partial y} + \frac{\partial E}{\partial x}\right), \tag{25}$$

$$g_2^2 = T_2^{-2}\left(\frac{\partial f}{\partial x} + \frac{\partial E}{\partial y}\right), \tag{26}$$

$$g_3^1 = -T_2^{-2}\nabla_1^{-2}\left(\nabla_1^2 w + \nabla^2 F\right), \tag{27}$$

$$g_3^2 = T_2^{-2}(F + w - E). \tag{28}$$

where $\nabla^2 E = 0, F = -\nabla_1^2 g_3^1 - \nabla_2^2 g_3^2 + E$.

In this way, the displacement can be derived as:

$$
\begin{aligned}
u_x = &\sum_{j=1}^{2} (-1)^{j-1}\frac{\sin(z\nabla_j)}{\nabla_j}\frac{\partial w}{\partial x} - \frac{\sin(z\nabla_2)}{\nabla_2}\left(\frac{\partial F}{\partial x} + \frac{\partial w}{\partial x}\right) \\
&- T_2^{-2}\sum_{j=1}^{2} (-1)^{j-1}\frac{\sin(z\nabla_j)}{\nabla_j}\nabla^2\left(\frac{\partial F}{\partial x} + \frac{\partial w}{\partial x}\right)
\end{aligned}
\tag{29}
$$

$$
\begin{aligned}
u_y = &\sum_{j=1}^{2} (-1)^{j-1}\frac{\sin(z\nabla_j)}{\nabla_j}\frac{\partial w}{\partial y} - \frac{\sin(z\nabla_2)}{\nabla_2}\left(\frac{\partial F}{\partial y} - \frac{\partial f}{\partial x}\right) \\
&- T_2^{-2}\sum_{j=1}^{2} (-1)^{j-1}\frac{\sin(z\nabla_j)}{\nabla_j}\nabla^2\left(\frac{\partial F}{\partial y} + \frac{\partial w}{\partial y}\right)
\end{aligned}
\tag{30}
$$

$$u_z = \cos(z\nabla_1)w - T_2^{-2}\sum_{j=1}^{2}(-1)^{j-1}\cos(z\nabla_2)\nabla^2(F + w). \tag{31}$$

In line with Hooke's law, the stress components can be expressed as:

$$
\begin{aligned}
\tau_{zx} = &2\mu\cos(z\nabla_1)\frac{\partial w}{\partial x} - \mu\cos(z\nabla_2)\left(\frac{\partial F}{\partial x} + \frac{\partial w}{\partial x} + \frac{\partial f}{\partial y}\right) \\
&- 2\mu T_2^{-2}\sum_{j=1}^{2}(-1)^{j-1}\cos(z\nabla_j)\nabla^2\left(\frac{\partial F}{\partial x} + \frac{\partial w}{\partial x}\right)
\end{aligned}
\tag{32}
$$

$$\tau_{zy} = 2\mu \cos(z\nabla_1)\frac{\partial w}{\partial y} - \mu \cos(z\nabla_2)\left(\frac{\partial F}{\partial y} + \frac{\partial w}{\partial y} - \frac{\partial f}{\partial x}\right)$$
$$-2\mu T_2^{-2}\sum_{j=1}^{2}(-1)^{j-1}\cos(z\nabla_j)\nabla^2\left(\frac{\partial F}{\partial y} + \frac{\partial w}{\partial y}\right) \tag{33}$$

$$\sigma_z = 2\mu T_2^{-2}\sum_{j=1}^{2}(-1)^{j-1}\frac{\sin(z\nabla_j)}{\nabla_j}\nabla^2\nabla^2(F+w)$$
$$-(\lambda+2\mu)T_1^{-2}T_2^{-2}\frac{\sin(z\nabla_1)}{\nabla_1}\nabla^2(F+w)$$
$$+(\lambda+2\mu)T_1^{2}\frac{\sin(z\nabla_1)}{\nabla_1}w \tag{34}$$
$$-2\mu\frac{\sin(z\nabla_1)}{\nabla_1}\nabla^2 w + 2\mu\frac{\sin(z\nabla_2)}{\nabla_2}\nabla^2(F+w)$$

Considering the free boundary condition, the shear stress at the plate surface is zero. The equalities can be given from Equations (32) and (33) as follows:

$$\left[\cos\left(\frac{h}{2}\nabla_1\right)\frac{\partial w}{\partial x} - \frac{1}{2}\cos\left(\frac{h}{2}\nabla_2\right) - T_2^{-2}\sum_{j=1}^{2}(-1)^{j-1}\cos\left(\frac{h}{2}\nabla_j\right)\nabla^2\right]$$
$$\left(\frac{\partial F}{\partial x} + \frac{\partial w}{\partial x}\right) - \frac{1}{2}\cos\left(\frac{h}{2}\nabla_2\right)\frac{\partial f}{\partial y} = 0 \tag{35}$$

$$\left[\cos\left(\frac{h}{2}\nabla_1\right)\frac{\partial w}{\partial y} - \frac{1}{2}\cos\left(\frac{h}{2}\nabla_2\right) - T_2^{-2}\sum_{j=1}^{2}(-1)^{j-1}\cos\left(\frac{h}{2}\nabla_j\right)\nabla^2\right]$$
$$\left(\frac{\partial F}{\partial y} + \frac{\partial w}{\partial y}\right) + \frac{1}{2}\cos\left(\frac{h}{2}\nabla_2\right)\frac{\partial f}{\partial x} = 0 \tag{36}$$

where h is the thickness of the plate.

On the basis of the complex variable function theory, Equations (35) and (36) can be regarded as the real part and imaginary part, which consist of a Riemann condition of the analytic function. Now, the non-homogeneous solution of the equation does not influence the solution of the stress state, so there can be:

$$\cos\left(\frac{h}{2}\nabla_1\right)w - \left[\frac{1}{2}\cos\left(\frac{h}{2}\nabla_2\right) + T_2^{-2}\sum_{j=1}^{2}(-1)^{j-1}\cos\left(\frac{h}{z}\nabla_j\right)\nabla^2\right](F+w) = 0, \tag{37}$$

$$\cos\left(\frac{h}{2}\nabla_2\right)f = 0. \tag{38}$$

According to the integral function theory, Equation (38) can be expanded in series as:

$$\prod_{m=1}^{\infty}\left[1 - \frac{h^2\nabla_2^2}{(2m-1)^2\pi^2}\right]f = 0. \tag{39}$$

By means of truncating the infinite series, the following second-order elastic wave equation is given as:

$$\nabla^2 f - \left(\frac{\pi^2}{h^2} + \frac{1}{c_2^2}\frac{\partial^2}{\partial t^2}\right)f = 0. \tag{40}$$

According to the free boundary condition that the normal stress at the plate surface is zero, the governing equation for the elastic wave of the plate can be derived from Equations (34) and (37) as:

$$T_2^{-2}\sum_{j=1}^{2}(-1)^{j-1}\cos(\frac{h}{2}\nabla_j)\nabla^2(F+w) + \frac{1}{2}\cos(\frac{h}{2}\nabla_2)(F+w) - \cos(\frac{h}{2}\nabla_2)w = 0, \tag{41}$$

$$T_2^{-2}\sum_{j=1}^{2}(-1)^{j-1}\frac{\sin(\frac{h}{2}\nabla_j)}{\nabla_j}\nabla^2\nabla^2(F+w) - \frac{1}{2}\left[\frac{\sin(\frac{h}{2}\nabla_1)}{\nabla_1} - \frac{2\sin(\frac{h}{2}\nabla_2)}{\nabla_2}\right]\nabla^2(F+w)$$
$$-\frac{\sin(\frac{h}{2}\nabla_1)}{\nabla_1}\nabla^2 w + \frac{1}{2}T_2^{2}\frac{\sin(\frac{h}{2}\nabla_1)}{\nabla_1}w = 0 \tag{42}$$

The operator algebraic equation involving the unknown functions F and w is deduced by Equations (41) and (42) as:

$$\Lambda \begin{bmatrix} F \\ w \end{bmatrix} = \begin{bmatrix} 0 \\ 0 \end{bmatrix}, \tag{43}$$

where the expression of each operator of Λ is

$$\Lambda_{11} = T_2^{-2} \sum_{j=1}^{2} (-1)^{j-1} \cos(\tfrac{h}{2}\nabla_j)\nabla^2 + \tfrac{1}{2}\cos(\tfrac{h}{2}\nabla_2),$$

$$\Lambda_{12} = T_2^{-2} \sum_{j=1}^{2} (-1)^{j-1} \cos(\tfrac{h}{2}\nabla_j)\nabla^2 - \cos(\tfrac{h}{2}\nabla_1) + \tfrac{1}{2}\cos(\tfrac{h}{2}\nabla_2),$$

$$\Lambda_{21} = T_2^{-2} \sum_{j=1}^{2} (-1)^{j-1} \frac{\sin(\tfrac{h}{2}\nabla_j)}{\nabla_j}\nabla^2\nabla^2 - \tfrac{1}{2}\frac{\sin(\tfrac{h}{2}\nabla_1)}{\nabla_1}\nabla^2 + \frac{\sin(\tfrac{h}{2}\nabla_2)}{\nabla_2}\nabla^2,$$

$$\Lambda_{22} = T_2^{-2} \sum_{j=1}^{2} (-1)^{j-1} \frac{\sin(\tfrac{h}{2}\nabla_j)}{\nabla_j}\nabla^2\nabla^2 - \tfrac{3}{2}\frac{\sin(\tfrac{h}{2}\nabla_1)}{\nabla_1}\nabla^2 + \frac{\sin(\tfrac{h}{2}\nabla_2)}{\nabla_2}\nabla^2 + \tfrac{1}{2}T_2^2\frac{\sin(\tfrac{h}{2}\nabla_1)}{\nabla_1}.$$

The determinant of Equation (23) is:

$$\begin{aligned} \det(\Lambda) = &\ T_2^{-2}\left[\frac{\sin(\tfrac{h}{2}\nabla_1)}{\nabla_1}\cos(\tfrac{h}{2}\nabla_2) - \frac{\sin(\tfrac{h}{2}\nabla_2)}{\nabla_2}\cos(\tfrac{h}{2}\nabla_1) \right]\nabla^2\nabla^2 \\ &- \left[\frac{\sin(\tfrac{h}{2}\nabla_1)}{\nabla_1}\cos(\tfrac{h}{2}\nabla_2) - \frac{\sin(\tfrac{h}{2}\nabla_2)}{\nabla_2}\cos(\tfrac{h}{2}\nabla_1) \right]\nabla^2 \\ &+ \tfrac{1}{4}T_2^2\frac{\sin(\tfrac{h}{2}\nabla_1)}{\nabla_1}\cos(\tfrac{h}{2}\nabla_2) \end{aligned} \tag{44}$$

The fourth-order differential equation involving the lateral displacement can be expressed as:

$$\det(\Lambda)w = 0. \tag{45}$$

After the truncation of the infinite order operator series, the governing equation for the elastic wave of plates can be derived as:

$$\nabla^2\nabla^2 w - \frac{3-2\kappa}{2(1-\kappa)}T_2^2\nabla^2 w + \frac{3}{1-\kappa}T_2^2\left(\frac{1}{h^2} + \frac{1}{24}T_1^2 + \frac{1}{8}T_2^2\right)w = 0, \tag{46}$$

$$C\nabla^2\nabla^2 w - (2-v)DT_2^2\nabla^2 w + \left[CT_2^2 + \left(\frac{7}{8}-v\right)DT_2^4\right]w = 0, \tag{47}$$

where $C = \frac{E_M h}{2(1+v)}$, $D = \frac{E_M h}{12(1-v^2)}$, E_M is Young modulus.

Without loss of generality, the solution of the vibration harmonic of the problem is studied. Set:

$$w = \tilde{w}e^{-i\omega t}, \quad F = \tilde{F}e^{-i\omega t}, \quad f = \tilde{f}e^{-i\omega t}, \tag{48}$$

where ω is the angular frequency of plate bending, and i is an imaginary unit.

In the following analysis, the time factor and the symbol '~' in the generalized displacement functions are left out. Taking Equation (48) into Equation (47), the following equations can be expressed as:

$$\nabla^2\nabla^2 w + \frac{3-2\kappa}{2(1-\kappa)}k_2^2\nabla^2 w + \frac{3}{4(1-\kappa)}k_2^2\left(\frac{\kappa k_2^2}{6} + \frac{k_2^2}{2} - \frac{4}{h^2}\right)w = 0, \tag{49}$$

$$\prod_{j=1}^{2}\left(\nabla^2 + \alpha_j^2\right)w = 0, \tag{50}$$

where $\alpha_j (j=1,2)$ are scattering wave numbers, which satisfy the following expression

$$\alpha^4 - (2-v)k_2^2\alpha^2 + k_2^2\left[\frac{(7-8v)k_2^2}{8} - \frac{6(1-v)}{h^2}\right] = 0.$$

The scattering numbers on the basis of Mindlin plate theory are determined by the following expression: $\alpha^4 - \left(\frac{12}{\pi^2} + \frac{1-v}{2}\right)k_2^2\alpha^2 + 6(1-v)k_2^2\left(\frac{k_2^2}{\pi^2} - \frac{1}{h^2}\right) = 0$, where $k_j = \frac{\omega}{c_j}$, $(j = 1,2)$.

The corresponding generalized displacement potential function is:

$$F = F_1 + F_2 = \sum_{j=1}^{2}(\delta_j - 1)w_j, \tag{51}$$

where δ_j are the ratio coefficients of the displacement potential function, and thus $\delta_j = \frac{16+2\left(\alpha_j^2-\kappa k_2^2\right)h^2}{8+\left[(3-2\kappa)\alpha_j^2-k_2^2\right]h^2}$.

Comparatively, the ratio coefficients of the displacement potential function by Mindlin plate theory are $\delta_1 = -\frac{2}{1-v}\frac{\alpha_2^2h^2}{\pi^2-k_2^2h^2}, \delta_2 = -\frac{2}{1-v}\frac{\alpha_1^2h^2}{\pi^2-k_2^2h^2}$.

3. Comparison of Various Bending Plate Theories

In this paper, the comparisons between the governing equation for the bending plate and the governing equations for various classical bending plates are presented in Table 1.

Table 1. Comparison of different theories of plate bending.

Plate Theory Categories	Statics Equations	Dynamics Equations
The exact plate theory in this paper	$DV^2V^2W = q - \frac{2-v}{8(1-v)}h^2\nabla^2q$ $\nabla^2f - \frac{\pi^2}{h^2}f = 0$	$D\nabla^2\nabla^2W - (2-v)DT_2^2\nabla^2W$ $+CT_2^2W + \left(\frac{7}{8}-v\right)W$ $= q + \frac{3D}{4C}(1-v)\left(T_2^2 - \frac{2-v}{1-v}\nabla^2\right)q$ $\nabla^2f - \left(\frac{\pi^2}{h^2}+T_2^2\right)f = 0$
Lagrange-Germain plate theory	$D\nabla^2\nabla^2W = q$	$D\nabla^2\nabla^2W + CT_2^2W = q$
Reissner plate theory	$D\nabla^2\nabla^2W = q - \frac{2-v}{10(1-v)}h^2\nabla^2q$ $\nabla^2f - \frac{10}{h^2}f = 0$	
Mindlin plate theory		$D\nabla^2\nabla^2W - \left(\frac{12\rho D}{\pi^2C} + \frac{\rho h^3}{12}\right)T_2^2\nabla^2W$ $+\rho hT_2^2W + \frac{\rho^2h^3}{\pi^2C}T_2^4W$ $= q + \frac{12}{\pi^2C}\left(\frac{\rho h^2}{12}T_2^2 - \frac{D}{h}\nabla^2\right)q$ $\nabla^2f - \left(\frac{3\pi^2}{4h^2}+T_2^2\right)f = 0$
Hencky plate theory	$D\nabla^2\nabla^2W = q - \frac{1}{6(1-v)}h^2\nabla^2q$	
Panc plate theory	$D\nabla^2\nabla^2W = q - \frac{1}{5(1-v)}h^2\nabla^2q$	

The dispersion equation based on the Lagrange–Germain plate theory (CPT) can be described as:

$$\frac{c}{c_2} = \sqrt{\frac{4\pi^2}{6(1-v)}\frac{h}{\lambda}}. \tag{52}$$

where λ is the wavenumber of an elastic wave.

The dispersion equation in consideration of the moment of inertia can be described as:

$$\frac{c}{c_2} = \sqrt{\frac{4\pi^2}{6(1-v)}\left[1 + \frac{\pi^2}{3}\left(\frac{h}{\lambda}\right)^2\right]^{-1}\frac{h}{\lambda}}. \tag{53}$$

When the moment of inertia and shear deformation are involved, the implicit dispersion equation can be described as:

$$\frac{\pi^2}{3}\left(\frac{h}{\lambda}\right)^2\left(1-\frac{c^2}{\kappa^2 c_2^2}\right)\left(\frac{c_p^2}{c^2}-1\right)=1. \tag{54}$$

The implicit dispersion equation based on the three-dimensional elasto-dynamics theory can be described as:

$$\frac{4c_2^2\sqrt{\left(c_2^2-\kappa c^2\right)\left(c_2^2-c^2\right)}}{\left(2c_2^2-c^2\right)^2}=\frac{\tanh\left(2\pi\frac{h}{\lambda}\sqrt{c_2^2-\kappa c^2}\right)}{\tanh\left(2\pi\frac{h}{\lambda}\sqrt{c_2^2-c^2}\right)}. \tag{55}$$

The dispersion equation based on the exact plate theory in this paper is:

$$\alpha^4-(2-v)k_2^2\alpha^2+(1-v)k_2^2\left[\frac{k_2^2(7-8v)}{8(1-v)}-\frac{6}{h^2}\right]=0. \tag{56}$$

4. Discussion of the Exact Plate Theory

In this paper, the derived plate bending vibration equation is compared with the classical corresponding equation. The comparison of the specific equation form is shown in Table 1. In the process of comparison, the equation form in the frequency domain is used. The bending vibration equation of plates presented in this paper is similar to other classical bending vibration equations of plates. When the statics problem is studied, the elastic vibration equation of plates derived in this paper degenerates into an exact equation of the static bending of plates.

According to those dispersion equations above, Figure 1 is drawn to compare the dispersion curves; Figure 2 is drawn to compare the scattering wave numbers. From Figure 1, we can see that the dispersion curves based on the classic thin plate theory and Mindlin plate theory are far apart from the three-dimensional elasto-dynamics theory, but the dispersion curves by the exact plate theory in this paper are very close to the dispersion curves based on the three-dimensional elasto-dynamics theory. By comparing those curves, the superiority of the exact plate theory to other plate theories is obvious.

As can be seen in Figure 2, the scattering wave number α_1 obtained by the exact plate theory and Mindlin plate theory is very close, but the scattering wave number α_2 obtained by the exact plate theory is quite different from that of the Mindlin plate theory. With the increase of the vibration frequency, the scattering wave number becomes greater. It can be seen that the scattering wave number α_1 at any frequency is greater than zero, so it can be concluded that the wave mode is in the propagation region. Nevertheless, when the frequency is low, the scattering wave number α_2 is less than zero, and the wave mode is in the cutoff frequency domain, which is called a localized standing wave. When the frequency is high, the scattering wave number α_2 is greater than zero, and the wave mode is in the propagation region, which is called a propagating wave. According to Reference [7], it can be seen that the applicable frequency interval of the Mindlin plate theory is $\omega/\omega_0 < 1$, i.e., $h/\lambda_2 < 0.5$, and the application of the Mindlin plate theory is limited. The dynamic model proposed in this paper is completely based on the three-dimensional elasto-dynamics theory, and consequently, its limitations of application are minor.

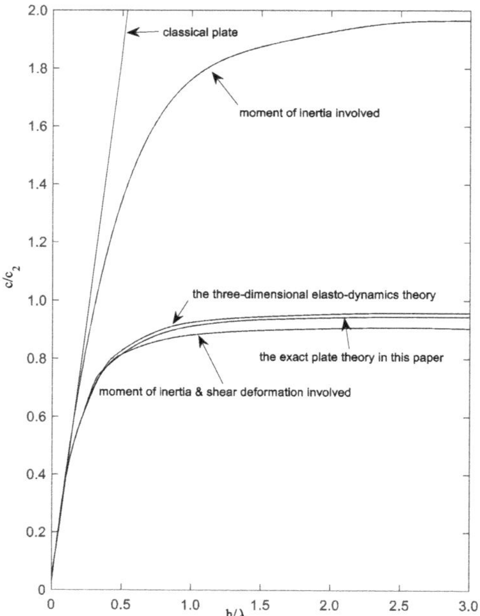

Figure 1. Dispersion relation by the different theories.

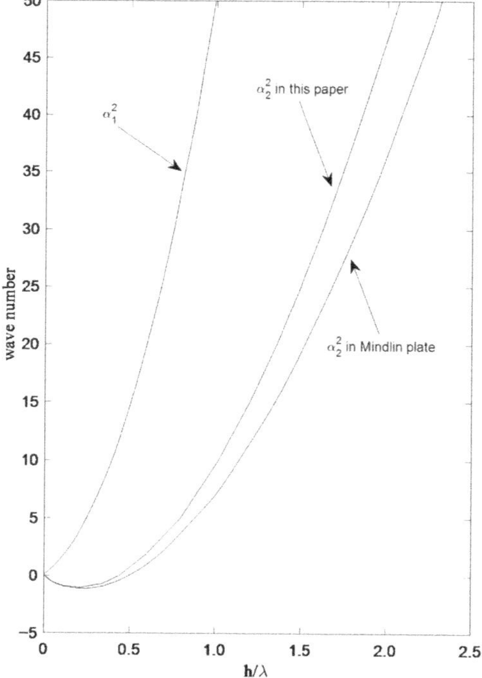

Figure 2. Wavenumbers by the different theories.

5. Conclusions

In this paper, the novel dynamics plate model is completely based on the theory of three-dimensional elasto-dynamics theory, and the derivation of the dynamic equation is conducted without any prior assumption but in view of the partial differential operator theory and analytic theory of complex function. Therefore, the proposed dynamic equation is more exact. It is not only suitable for low-frequency vibration of the plate but also suitable for high-frequency vibration, which will not produce the phenomenon of high-frequency dispersion.

With the development of modern mechanics and mathematics, new research results and methods continue to appear. The research object develops toward high speed and high-frequency. The new development of mathematical and mechanics improves the ability to solve complex mechanical problems and plays a positive role in promoting the development of mathematics in mechanics and physics. It provides a new formulation and means for solving the dynamics and numerical calculation of solid structures more accurately. The governing equation of bending plate vibration proposed in this paper is expected to be used to analyze the vibration of thick plates, evaluate the applicable condition of the engineering plate theory and design the active vibration control of thick plates exactly in the research on the dynamics and stability for flexible spacecraft structures and broadband vibration frequency.

Author Contributions: Conceptualization, C.Z. and M.W.; validation, C.Z. and M.W.; investigation, X.H. and W.Z.; data curation, X.H. and H.X.; writing—original draft preparation, C.Z.; writing—review and editing, M.W. and J.N.; funding acquisition, J.N. All authors have read and agreed to the published version of the manuscript.

Funding: This research was supported by the Key Laboratory of Underwater Acoustic Environment Institute of Acoustic Chinese Academy of Science SSHJ-KFKT-2020, the Key Research and Development Program of Zhejiang Province (No. 2021C03013), the National Key Research & Development Program of China (No. 2017YFB1301300), the Key Laboratory funding for Technology in Rural Water Management of Zhejiang Province and the Fundamental Research Funds for the Provincial Universities of Zhejiang (GK199900299012-026).

Institutional Review Board Statement: The study in this paper did not involve humans or animals.

Informed Consent Statement: Not applicable.

Data Availability Statement: The study did not report any data.

Conflicts of Interest: The authors declare no conflict of interest.

References

1. Corral, E.; García, M.J.G.; Castejon, C.; Meneses, J.; Gismeros, R. Dynamic modeling of the dissipative contact and friction forces of a passive biped-walking robot. *J. Appl. Sci.* **2020**, *10*, 2342. [CrossRef]
2. Corral, E.; Moreno, R.G.; García, M.J.G.; Castejón, C. Nonlinear phenomena of contact in multibody systems dynamics: A review. *Nonlinear Dyn.* **2021**, *104*, 1269–1295. [CrossRef]
3. Kolahchi, R. A comparative study on the bending, vibration and buckling of viscoelastic sandwich nano-plates based on dif-ferent nonlocal theories using DC, HDQ and DQ methods—ScienceDirect. *J. Aerosp. Sci. Technol.* **2017**, *66*, 235–248. [CrossRef]
4. Nicassic, F. Shape prediction of bistable plates based on Timoshenko and Ashwell theories. *Compos. Struct.* **2021**, *265*, 113645. [CrossRef]
5. Wu, C.; Viquerat, A.; Aglietti, G. Natural Frequency Optimization and Stability Analysis of Bistable Carbon Fiber Reinforced Plastic Booms for Space Applications. In Proceedings of the 3rd AIAA Spacecraft Structures Conference, San Diego, CA, USA, 4–8 January 2016. [CrossRef]
6. Saada, A.S. *Elasticity: Theory and Applications*; J. Ross Pub.: Ft. Lauderdale, FL, USA, 2009.
7. Pao, Y.H.; Mow, C.C. Diffraction of elastic wave and dynamic stress concentration. *J. Appl. Math.* **1973**, 872. [CrossRef]
8. Shimpi, R.P.; Shetty, R.A.; Guha, A. A single variable refined theory for free vibrations of a plate using inertia related terms in displacements. *Eur. J. Mech. A-Solid* **2017**, *65*, 136–148. [CrossRef]
9. Wang, C.M.; Kitipornchai, S. Frequency relationship between Levinson plates and classical thin plates. *Mech. Res. Commun.* **1999**, *26*, 687–692. [CrossRef]

10. Wang, C.M.; Reddy, J.N.; Lee, K.H. (Eds.) *Shear Deformable Beams and Plates: Relationships with Classical Solutions*, 1st ed.; Elsevier: Amsterdam, The Netherlands, 2000.
11. Reissner, E. On the Theory of Bending of Elastic Plates. *J. Math. Phys.* **1944**, *23*, 184–191. [CrossRef]
12. Reissner, E. The Effect of Transverse Shear Deformation on the Bending of Elastic Plates. *J. Appl. Mech.* **1945**, *12*, A69–A77. [CrossRef]
13. Mindlin, R.D. Influence of rotary inertia and shear on flexural motions of isotropic elastic plates. *J. Appl. Mech.* **1951**, *18*, 31–38. [CrossRef]
14. Reissner, E. Reflections on the Theory of Elastic Plates. *Appl. Mech. Rev.* **1985**, *38*, 1453–1464. [CrossRef]
15. Levinson, M. An accurate, simple theory of the statics and dynamics of elastic plates. *Mech. Res. Commun.* **1980**, *7*, 343–350. [CrossRef]
16. Reddy, J.N. A Simple Higher-Order Theory for Laminated Composite Plates. *J. Appl. Mech.* **1984**, *51*, 745–752. [CrossRef]
17. Reddy, J. A refined nonlinear theory of plates with transverse shear deformation. *Int. J. Solids Struct.* **1984**, *20*, 881–896. [CrossRef]
18. Kant, T. Numerical analysis of thick plates. *Comput. Methods Appl. Mech. Eng.* **1982**, *31*, 1–18. [CrossRef]
19. Wang, X.; Shi, G. A refined laminated plate theory accounting for the third-order shear deformation and interlaminar transverse stress continuity. *Appl. Math. Model.* **2015**, *39*, 5659–5680. [CrossRef]
20. Nedri, K.; Meiche, N.; Tounsi, A. Free vibration analysis of laminated composite plates resting on elastic foundations by using a refined hyperbolic shear deformation theory. *J. Mech Compos Mater.* **2014**, *49*, 629–640. [CrossRef]
21. Shimpi, R.P. Refined plate theory and its variants. *AIAA J.* **2002**, *40*, 137–146. [CrossRef]
22. Shimpi, R.P.; Patel, H.G. Free vibrations of plate using two variable refined plate theory. *J. Sound Vibr.* **2006**, *296*, 979–999. [CrossRef]
23. Carrera, E.; Filippi, M.; Zappino, E. Free vibration analysis of rotating composite blades via Carrera Unified Formulation. *Compos. Struct.* **2013**, *106*, 317–325. [CrossRef]
24. Tornabene, F.; Viola, E.; Fantuzzi, N. General higher-order equivalent single layer theory for free vibrations of doubly-curved laminated composite shells and panels. *Compos. Struct.* **2013**, *104*, 94–117. [CrossRef]
25. Tornabene, F.; Fantuzzi, N.; Viola, E.; Carrera, E. Static analysis of doubly-curved anisotropic shells and panels using CUF approach, differential geometry and differential quadrature method. *Compos. Struct.* **2014**, *107*, 675–697. [CrossRef]

 *mathematics*

Article

Power Grid Dynamic Performance Enhancement via STATCOM Data-Driven Control

David Rivera [1], Daniel Guillen [1,*], Jonathan C. Mayo-Maldonado [2], Jesus E. Valdez-Resendiz [1] and Gerardo Escobar [1]

[1] School of Engineering and Sciences, Tecnologico de Monterrey, Monterrey 64849, Mexico; A00819594@itesm.mx (D.R.); jesusvaldez@tec.mx (J.E.V.-R.); gerardo.escobar@tec.mx (G.E.)
[2] Department of Electronic and Electrical Engineering, The University of Sheffield, Sheffield S10 2TN, UK; j.mayo@sheffield.ac.uk
* Correspondence: guillenad@tec.mx

Abstract: This work proposes a data-driven approach to controlling the alternating current (AC) voltage via a static synchronous compensator (STATCOM). This device offers a fast dynamic response injecting reactive power to compensate the voltage profile, not only during load variations but also depending on the operating point established by the grid. The proposed control scheme is designed to improve the dynamic grid performance according to the defined operating point into the grid. The mathematical fundamentals of the proposed control strategy are described according to a (model-free) data-driven-based controller. The robustness of the proposed scheme is proven with several tests carried out using Matlab/Simulink software. The analysis is performed with the well-known test power system of two areas, demonstrating that the proposed controller can enhance the dynamic performance under transient scenarios. As the main strength of the present work with respect to the current state-of-the-art, we highlight the fact that no prior knowledge of the system is required for the controller implementation, that is, a model or a system representation. The synthesis of the controller is obtained in a pure numerical way from data, while it can simultaneously ensure stability in a rigorous way, by satisfying Lyapunov conditions.

Keywords: data-driven control; reactive power compensation; STATCOM; voltage control; voltage source converter

Citation: Rivera. D.; Guillen, D.; Mayo-Maldonado, J.C., Valdez-Resendiz, J.E.; Escobar, G. Power Grid Dynamic Performance Enhancement via STATCOM Data-Driven Control. *Mathematics* **2021**, *9*, 2361. https://doi.org/10.3390/math9192361

Academic Editor: Roberto Salvado-Félix Patrón

Received: 22 July 2021
Accepted: 1 September 2021
Published: 23 September 2021

Publisher's Note: MDPI stays neutral with regard to jurisdictional claims in published maps and institutional affiliations.

1. Introduction

The technical regulations about environmental issues and the use of renewable energy sources (RES) set robust planning expansion programs due to the increasing energy demand. This leads to analysis of the system constraints, aiming to avoid instability scenarios defined by the load-ability limits [1]. Voltage stability is of utmost importance in electrical power systems studies and is related to reactive power compensation. In this context, flexible AC transmission system (FACTS) devices have been developed not only to offer a fast dynamic response of reactive power compensation but also to prevent the occurrence of synchronous resonance in large power systems [2,3]. In the literature, it has been reported that static synchronous compensator (STATCOM) presents promising results in dynamic reactive power compensation, voltage regulation and also helping to reduce power fluctuations [4,5].

During transient events, a power system may exhibit low-frequency power oscillation between two or more interconnected areas, also called inter-area oscillations, that are in a range of 0.1–1.0 Hz according to [6]. It is well-known that a STATCOM model not only provides good support of voltage control but can also improve the dynamic performance of the grid during transient events, in other words, is able to enhance the voltage recovery time and limits, and it is expected to have a better performance during power oscillations. As a consequence, a STATCOM is a good alternative to provide power oscillation damping

(POD) [7]. This task is commonly carried out using power system stabilizers (PSS) to damp low-frequency power oscillations, which consists of an auxiliary control loop that measures the deviation of frequency. However, this is not the main task of a STATCOM, it is rather an inherent response due to its control design and can also include an additional damping controller [8].

Dynamic reactive power compensation can be an effective way to facilitate the interconnection of RES, especially to comply with grid codes regarding reactive power requirements at the point of common coupling (PCC) [9]. For example, in [10], a control scheme based on the fundamentals of a STATCOM is employed in photovoltaic (PV) power plants to reduce power oscillations in power grids, due to an inverter being able to act as a STATCOM; this enables the mitigation of POD during transient events. In [11], a STATCOM is also used in wind power plants to enhance the dynamic performance of the main oscillating modes; in that work, an adaptive network-based fuzzy inference system (ANFIS) controller is employed. Another proposal focused on POD can be found in [12], where a STATCOM is equipped with energy storage so that the combination of real and reactive power injection offers good robustness during the dynamic response of the power system.

On the other hand, different control strategies have been proposed in the literature to enhance the dynamic response of power systems supported by STATCOMs. Many of those approaches present different attributes according to the employed techniques. In [13], a mechanism of POD is based on auxiliary damping controllers defined by a wavelet neural network (WNN), which offers a reduced complexity due to the number of data used as well as its learning capability. In the same way, in [14], a multi-band controller is employed to deal with POD, which is a coordinated design and is optimized based on the operating conditions of the grid. Most of them have been used to design auxiliary control loops aiming to damp power oscillations, like in [8], which presents a control strategy using an additional damper controller (ADC) based on artificial neural networks and deep deterministic policy gradient (DDPG). In another study reported in [15], a data-driven analysis is carried out to adjust and calculate the control gain to enhance the overall system dynamic response to reduce the power oscillations. In contrast with other proposed control strategies, this work is focused on the full control of the voltage source converter (VSC)-based STATCOM without using auxiliary control loops to enhance the dynamic performance.

This paper presents a fully data-driven controller for a VSC-based STATCOM. The general idea is based on the fact that a STATCOM helps to increase the transient stability; this is feasible using well-coordinated controllers or employing other supplementary control functions. In this context, the proposed approach aims to improve the dynamic performance of a VSC-based STATCOM following the fundamentals of a conventional VSC controller. The contribution is underpinned in a data-driven controller to enhance the transient response of a VSC-based STATCOM under different operating conditions of a two area, four-machine power system. The proposed controller offers robustness and adaptability according to the power system requirements that will be reflected in the dynamic performance.

The proposed data-driven control is able to ensure the demanding performance specifications without any prior knowledge of the system. Among such specifications we can highlight stability as the most important one. It is well-known that model-based techniques are able to ensure stability and general performance. However, the limitation of such approach is the requirement of an existing accurate model of the grid and the power converter as a starting point. For instance, in [16] the small signal model of the system is required to design the controller of a STATCOM based on a multilevel converter. Similarly, in [17] the model of the system is required in order to design an adaptive controller for voltage regulation using a STATCOM. Other newer approaches, such as model predictive control (MPC), also require a detailed modeling of the system in order to work properly as can be corroborated in [18,19], where MPC is used as the control strategy for the mitigation of voltage unbalance and reactive power control. Robust control can be also categorized as a model-based technique as can be observed in [20,21], where the system-model derivation is an important step of the controller design procedure. In real-life, a power grid is such a complex system that its model, parameters and general dynamics have a high-level of uncertainty. Consequently, the stability conditions obtained from an idealized model can be compromised during extreme scenarios such as the occurrence of faults. Another alternative is the use of classical proportional (P), proportional and integral (PI) and proportional, integral and derivative (PID) tuning rules, which do not require a model to set-up controller gains. The limitation of these rules, however, is the fact that they cannot guarantee stability unless a system model, for example, transfer function or frequency response traces, is provided (see e.g., [22,23]). Motivated by this problem, we developed a data-driven control technique that is able to ensure performance specifications and by all means stability, as in a model-based technique. Nevertheless, we replace the requirement of a model, by matrices constructed from data and stability conditions provided in terms of linear matrix inequalities (LMIs), which can be easily set-up and numerically solved by traditional MATLAB (Version R2021a, MathWorks, Natick, MA, USA) toolboxes such as Yalmip (free toolbox developed by Dr. Johan Lofberg).

The present paper is organized as follows. In Section 2, a theoretical background is discussed, which sets the basis of the proposed data-driven controller. The conventional state-space model is replaced by data, which is discussed in Section 3; it includes modeling and parametric identification. Section 4 describes the data-driven control design. The conventional and data-driven VSC is derived in Section 5, which includes the development of the state-space model, also known as the model-based. The case of study is depicted in Section 6, which describes the system that is going to be used to test the data-driven controller. In addition, Section 6 also includes the results, and the tests are focused on three aspects: voltage reference changes, power oscillation damping during transient faults, and load shedding. The data-driven controller performance is compared to the conventional controller, which is a state-space model described in [24]. Finally, Section 7 presents the conclusion of the data-driven controller performance.

2. Theoretical Background

In this section, we introduce the main notation and theoretical elements that constitute the basis of the proposed data-driven controller.

2.1. Notation

The notation used throughout the paper is described next. $\mathbb{R}$ is the set of reals, and $\mathbb{Z}_+$ is the set of positive integers. $\mathbb{R}^q$ stands for real vectors of dimension q. $\mathbb{R}^{p \times q}$ represents real matrices of dimension $p \times q$. An identity matrix with q rows and q columns is denoted by I_q. $\mathrm{col}(x_1, x_2)$ is a vector obtained after stacking column vectors x_1 over x_2. $\mathrm{rank}(M)$ denotes the rank of matrix $M \in \mathbb{R}^{p \times q}$, and $\mathrm{colspan}(M)$ represents the set of all linear combinations of its column vectors. σ denotes the shift operator, which applies to a function $f : \mathbb{Z}_+ \to \mathbb{R}^q$ in the form $(\sigma f)(t) := f(t+1)$. This operator can be extended to an order N, as $(\sigma^N f)(t) := f(t+N)$.

2.2. Linear Difference Systems

Recall that linear difference equations can be used to study discrete-time linear (sampled) systems, which have the following quite general form:

$$R_0 w + R_1(\sigma w) + \cdots + R_N(\sigma^N w) = 0, \tag{1}$$

where the discrete time function $w : \mathbb{Z}_+ \to \mathbb{R}^q$ maps time instants into physical amounts (or measurements); the maximum degree of the shift operator σ is represented by N; and $R_i \in \mathbb{R}^{p \times q}$ ($i = 0, 1, \ldots, N$). The linear difference system (1) can be compactly expressed as:

$$R(\sigma)w = 0 \; ; \tag{2}$$

where $R(\sigma)$ is a $p \times q$ polynomial matrix in σ, and represents the laws of the physical system with respect to w. The components of w can be classified as either inputs or outputs. Input functions, denoted by u, are independent (v.g., control variables), and output functions, denoted by y, are results due to the inputs (v.g., state variables). These variables can be accommodated as an *input/output partition*, that is, $w := \mathrm{col}(u, y)$.

2.3. Quadratic Difference Forms (QdFs)

Functionals, such as Lyapunov functions, have been traditionally used to study stability and other important properties of linear difference systems. In the present case, we use the notion of quadratic difference forms (QdFs), which are functionals of the discrete-time function w and its time-shifts, that is,

$$Q_{\Psi}(w) = \begin{bmatrix} w^{\top} & \sigma w^{\top} & \cdots & \sigma^N w^{\top} \end{bmatrix} \widetilde{\Psi} \begin{bmatrix} w \\ \sigma w \\ \vdots \\ \sigma^N w \end{bmatrix}, \tag{3}$$

where $\widetilde{\Psi} \in \mathbb{R}^{Nq \times Nq}$ is referred to as the coefficient matrix of Q_{Ψ}. The rate of change of functional Q_{Ψ}, denoted as ∇Q_{Ψ} (an analogous to a continuous-time derivative), is given by

$$\nabla Q_{\Psi}(w)(t) := \sigma Q_{\Psi}(w)(t) - Q_{\Psi}(w)(t) . \tag{4}$$

Stability for autonomous systems represented by (2) can thus be studied by means of QdFs. A system is autonomous if the polynomial matrix $R(\sigma)$ in (2) is square and nonsingular (see [25]). In the present case, we will see that this characteristic is easily achieved since the resultant closed-loop system under study is autonomous. An autonomous linear difference system is asymptotically stable if

$$\lim_{t \to \infty} w(t) = 0 , \quad \forall w \text{ satisfying (2)}.$$

A system described by (2) is asymptotically stable, according to the Lyapunov approach, if a QdF Q_{Ψ} exists and is such that, $\forall w$ satisfying (2), the following holds:
(i) $Q_{\Psi}(w) \geq 0$; and
(ii) $\nabla Q_{\Psi}(w) < 0$.

This QdF Q_{Ψ} that satisfies the above inequalities is referred to as the *Lyapunov function*.

2.4. Stabilization

We are now interested in designing a controller that is not only able to regulate the system variables to a desired set-point, but that can also guarantee stability during disturbances and events that are typical in an electrical system.

In terms of linear difference systems, the equations of the plant and the controller can be represented as in (2), that is, by $P(\sigma)w = 0$ and $C(\sigma)w = 0$, respectively. Moreover, the interconnected (closed-loop) system can be represented by:

$$\underbrace{\begin{bmatrix} P(\sigma) \\ C(\sigma) \end{bmatrix}}_{R(\sigma)} w = 0 , \tag{5}$$

where plant $P(\sigma)w = 0$ and controller $C(\sigma)w = 0$ laws must be simultaneously satisfied by w. This means that, by selecting a suitable controller, we are able to restrict the trajectories of the system to those that are asymptotically stable and discard those that are undesirable, for example, unstable, highly oscillatory, too slow, and so forth.

The design of controller $C(\sigma)$ can impose the stability on (5). For this, it must be guaranteed that, having a partition $w = \mathrm{col}(u, y)$, the stability conditions recalled in

Section 2.3 for a Lyapunov function candidate Q_Ψ, hold for all w satisfying (5). Notice that, if the coefficient matrix satisfies $\widetilde{\Psi} > 0$, then $Q_\Psi \geq 0$ prevails. Then it is still necessary to guarantee that $\nabla Q_\Psi < 0 \ \forall w$ satisfying (5). For this, the description of the closed-loop system can be introduced in the inequality, by considering a polynomial matrix $V(\sigma)$, which is non zero, and has the same dimensions as $R(\sigma)$ in (5). In this form, the symmetry necessary to satisfy the inequality is preserved, that is,

$$\underbrace{\sigma Q_\Psi(w) - Q_\Psi(w)}_{\nabla Q_\Psi(w)} + \underbrace{w^\top V(\sigma)^\top \begin{bmatrix} P(\sigma) \\ C(\sigma) \end{bmatrix} w + w^\top \begin{bmatrix} P(\sigma) \\ C(\sigma) \end{bmatrix}^\top V(\sigma) w}_{\text{Symmetric component}} < 0 \,. \tag{6}$$

Notice that the condition imposed by inequality (6) is interpreted as follows. If a QdF $Q_\Psi \geq 0$ (i.e., $\widetilde{\Psi} > 0$) exists and is such that (6) is satisfied, then asymptotic stability is guaranteed for the interconnected system (5). This follows from the fact that every trajectory w satisfying the interconnected system laws, will cancel out the additional symmetric component (because $P(\sigma)w = C(\sigma)w = 0$), which meets the condition $\nabla Q_\Psi < 0$, concerning the trajectories w produced by the closed-loop system (5).

Next, we introduce a numerical solution to this apparently algebraically complex condition. For this, we use a candidate controller whose gains are unknown that will be eventually computed using measurement data, rather than a model of the system. The plant mathematical model will be ultimately substituted by a condition on coefficient matrices built entirely from data.

3. Bypassing Models Using Data

The main objective of control design based on the data-driven approach is enabling the possibility to synthesize controllers entirely from data of available measurements, which is a simplified route with respect to the classic system identification (modeling and parametric identification) plus a model-based control approach. This is illustrated in Figure 1.

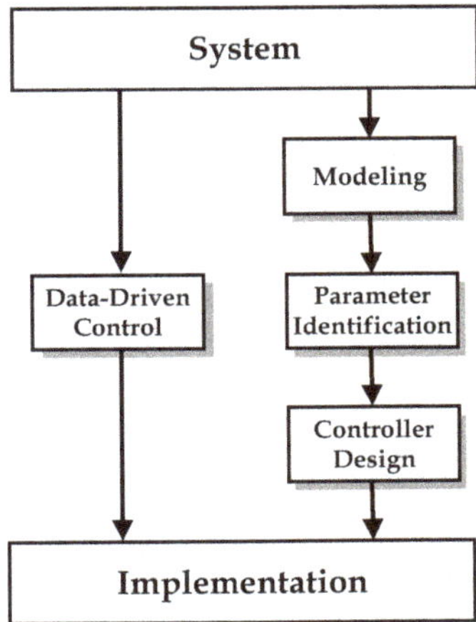

Figure 1. Proposed data-driven control vs. traditional "system identification + model-based" approach.

Moreover, stability conditions and the desired performance must be completely comparable with a model-based scheme. This is also in sharp contrast with basic empiric gain-tuning rules for classic P, PI, and PID controllers, which do not require a model of the system, but cannot guarantee stability and performance in a deterministic way (see [22,23]). Our present conviction is to generate a controller purely from measurements that permits to omit the need for an explicit mathematical model without losing stability and general performance capabilities with respect to a model-based technique.

The process begins with the establishment of essential conditions to assure that the measured data are convenient for control design. In brief, we introduce a test that determines whether the information provided by the available data is sufficient to fully recuperate the system physical laws.

3.1. Information Sufficiency

Consider that the sampled external variables w are aligned in a vector of measurement data $\{w(1), w(2), \ldots, w(T)\}$ of length T. Associated with this set of measurements, a *Hankel matrix* with a depth of $L < T \in \mathbb{Z}_+$ can be defined as follows:

$$\mathcal{H}_L(w) := \begin{bmatrix} w(t) & \sigma w(t) & \cdots & \sigma^{(T-L+1)}w(t) \\ \sigma w(t) & \sigma^2 w(t) & \cdots & \sigma^{(T-L+2)}w(t) \\ \vdots & \vdots & \cdots & \vdots \\ \sigma^L w(t) & \sigma^{(L+1)}w(t) & \cdots & \sigma^T w(t) \end{bmatrix}. \tag{7}$$

Next, the persistency of excitation concept [26] is appealed to; to verify if the available information provided by measurements is sufficient to recuperate the system physical laws, we use. This is a condition that applies for the input functions u in $w = \mathrm{col}(u, y)$, which is defined as follows. A vector $u = u(1), u(2), \ldots, u(T)$ is said to be persistently exciting (PE) of order L if matrix $\mathcal{H}_L(u)$ has full row rank.

Assume that u is PE of at least order L, where L equals the sum of the number of inputs plus the state-space dimension (please check Theorem 1 in [26]); out of this, $\mathrm{colspan}(\mathcal{H}_L(w))$ represents the set of all possible solutions of (2). That is, if the input is PE, then the complete dynamics of the electrical system can be fully described by the set of available measurements. While a model is able to determine all the possible outcomes of an electrical system as the solution of linear difference or differential equations, we are able to do the same by considering the linear combination of the row vectors of the Hankel matrix $\mathcal{H}_L$. Consequently, this array of data in a matrix owns the same model information.

3.2. Data-Based Coefficient Matrices

In this section we show the connection between matrices constructed from data and the models studied in Section 2.

Consider the above kernel representation (2). Based on (1), the following factorization is obtained:

$$\underbrace{\begin{bmatrix} P_0 & P_1 & \cdots & P_N \end{bmatrix}}_{\widetilde{P}} \begin{bmatrix} w \\ \sigma w \\ \vdots \\ \sigma^N w \end{bmatrix} = 0 \; ; \tag{8}$$

where $\widetilde{P}$ (a block matrix) is referred to as the coefficient matrix. It is shown next that this matrix can be directly obtained out of measured data. For this, consider expression (7) for $L = N + 1$, w with sufficiency of information and N representing the maximum degree of the shift operator. We appeal now to the singular-value decomposition (SVD) which is defined as $\mathcal{H}_{N+1}(w) := U \Sigma V^\top$, where matrices U and V are square ($q \times q$) and orthogonal; Σ represents a diagonal matrix having non-negative real numbers on its diagonal which are referred to as singular values. Furthermore, there is a number given by $r := \mathrm{rank}(\mathcal{H}_{N+1}(w))$ that accounts for non-zero singular values. It can be demonstrated, out of the last $(q - r)$ rows of zeros of $U^\top H(w)$, that $U^\top$ has annihilators of w (i.e., the set of vectors $\mathcal{V}$ such that $\mathcal{V}w = 0$). Therefore, after examining the partition $U := \begin{bmatrix} U_1 & U_2 \end{bmatrix}$, where U_1 owns r columns, the left kernel $\widetilde{P} := U_2^\top$ can be retrieved.

Notice that matrix $U_2^\top$, built entirely from data, owns the same information as that offered by the coefficient matrix $\widetilde{P}$, which is derived from an explicit mathematical model. Based on this proven equivalence, it is possible to get around the need for an explicit mathematical model representation and to directly design controllers considering only measured data and assisted by numerical tools.

4. Data-Based (Model-Free) Control

This section introduces a method to design stabilizing controllers from measured data. The proposed method involves the calculation of linear matrix inequalities (LMIs). For this, consider first that the elements of (6) can be factored in terms of coefficient matrices as described next. Notice that the energy rate of change can be factored as follows:

$$
\nabla Q_\Psi(w) = \underbrace{\begin{bmatrix} w \\ \sigma w \\ \vdots \\ \sigma^N w \end{bmatrix}^\top \begin{bmatrix} 0_{q\times q} & 0_{q\times Nq} \\ 0_{Nq\times q} & \widetilde{\Psi} \end{bmatrix} \begin{bmatrix} w \\ \sigma w \\ \vdots \\ \sigma^N w \end{bmatrix}}_{\sigma Q_\Psi(w)} - \underbrace{\begin{bmatrix} w \\ \sigma w \\ \vdots \\ \sigma^N w \end{bmatrix}^\top \begin{bmatrix} \widetilde{\Psi} & 0_{Nq\times q} \\ 0_{q\times Nq} & 0_{q\times q} \end{bmatrix} \begin{bmatrix} w \\ \sigma w \\ \vdots \\ \sigma^N w \end{bmatrix}}_{Q_\Psi(w)} . \tag{9}
$$

Based on the coefficient matrix definition for the plant dynamics presented in (8), and defining coefficient matrices for $C(\sigma)$ and $V(\sigma)$, the following factorizations can be obtained

$$
\begin{bmatrix} P(\sigma) \\ C(\sigma) \end{bmatrix} w = \begin{bmatrix} \widetilde{P} \\ \widetilde{C} \end{bmatrix} \begin{bmatrix} w \\ \sigma w \\ \vdots \\ \sigma^N w \end{bmatrix} , \quad V(\sigma)w = \widetilde{V} \begin{bmatrix} w \\ \sigma w \\ \vdots \\ \sigma^N w \end{bmatrix} . \tag{10}
$$

Notice that, out of factorizations (9) and (10), condition (6) can be entirely written in terms of coefficient matrices, that is,

$$
\begin{bmatrix} 0_{q\times q} & 0_{q\times Nq} \\ 0_{Nq\times q} & \widetilde{\Psi} \end{bmatrix} + \begin{bmatrix} \widetilde{\Psi} & 0_{Nq\times q} \\ 0_{q\times Nq} & 0_{q\times q} \end{bmatrix} + \widetilde{V}^\top \begin{bmatrix} \widetilde{P} \\ \widetilde{C} \end{bmatrix} + \begin{bmatrix} \widetilde{P}^\top & \widetilde{C}^\top \end{bmatrix} \widetilde{V} \geq 0 ; \tag{11}
$$

Consequently, if there is a $\widetilde{\Psi} = \widetilde{\Psi}^\top \geq 0$, $\widetilde{X} \in \mathbb{R}^{(N+1)q\times(N+1)q}$ and $\widetilde{C} \in \mathbb{R}^{(q-m)\times(N+1)q}$ such that (11) is kept valid, then stability is guaranteed for a plant whose coefficient matrix $\widetilde{P} \in \mathbb{R}^{(q-m)\times(N+1)q}$ is built upon data. It is noteworthy that the numerical solution of the inequality (11) is a relatively simple issue for conventional MATLAB toolboxes such as `Yalmip`. Therefore, based solely on measurement data to generate $\widetilde{P}$, the coefficients of a stabilizing controller can be obtained without the need for an explicit mathematical model. In other words, the controller given by $C(\sigma)w = 0$ can be realized out of the numerical solution of $\widetilde{C}$ in (11).

Candidate Controller for Stabilization

After examining (11), one can conclude that there are several solutions that will deliver convenient stabilization controllers for a certain plant. However, regarding electric power systems, there might be a particular interest in finding solutions that exhibit particular requirements, for instance, the regulation of certain variables despite of disturbances. As an example, in this section, a general convenient controller structure is proposed. The associated gains can be implicit in $\widetilde{C}$, and thus they can be numerically calculated after solving (11).

The controller design process starts by considering the error variables $\Delta x := x - x^*$, where x represents the original discrete-time function, while x^* is the reference at the equilibrium point (set point). Next, the following proportional feedback current control is proposed:

$$
\Delta u := -K\Delta y , \tag{12}
$$

where $\Delta u := u - u^*$ and $\Delta y := y - y^*$ are the error variables of the input u and the output y, respectively. We will denote the number of inputs as l and number of outputs as m; consequently $K \in \mathbb{R}^{m \times l}$.

This control loop can guarantee stabilization by a proper computation of K. Moreover, to ensure steady-state error compensation we can add a discrete-time integrator:

$$\Delta u = -K\Delta y - Gz \; ; \quad \sigma z = z + \Delta y \; ; \tag{13}$$

where z represents an auxiliary state-variable to describe the discrete-time integrator of the output variable error, then $G \in \mathbb{R}^{m \times l}$. By considering $w := \mathrm{col}(\Delta u, \Delta y, z)$ and considering (12) and (13), it is possible to obtain the following representation for the controller:

$$\underbrace{\begin{bmatrix} I_m & K & G \\ 0_{m \times l} & -I_l & \sigma I_m - I_m \end{bmatrix}}_{C(\sigma)} \begin{bmatrix} \Delta u \\ \Delta y \\ z \end{bmatrix} = 0 \; . \tag{14}$$

The associated coefficient matrix $\widetilde{C}$ is described by:

$$\widetilde{C} = \begin{bmatrix} I_m & K & G & 0_{m \times m} & 0_{m \times l} & 0_{m \times l} \\ 0_{l \times m} & -I_l & -I_l & 0_{l \times m} & -I_l & \sigma I_l \end{bmatrix} . \tag{15}$$

Now the gains in K and G can be numerically computed as a solution of (11) with $\widetilde{P}$ the coefficient matrix of the plant, which is obtained out of measured data w, as explained in Section 3.2.

5. Voltage Source Converters

Several VSC models have been developed and used throughout the past few years. Depending on the approach, a detailed or average model can be employed. The control design of the converter may vary and some designs are more accurate than others, but for large power systems an average model is usually used to reduce the computational effort [27].

5.1. Conventional VSC

The conventional VSC is also referred to as an average model, which is represented in Figure 2. This model consists of one controlled voltage source on the AC side, and another controlled current source on the direct current (DC) side. The state space model of the conventional VSC converter is derived by the connection to the grid through a power transformer; therefore, its control can be performed by the impedance Z_f among two buses v_g and v_c, defined by L_f and R_f. The equivalent circuit of the grid is represented by a grid impedance Z_s and a voltage source V_s.

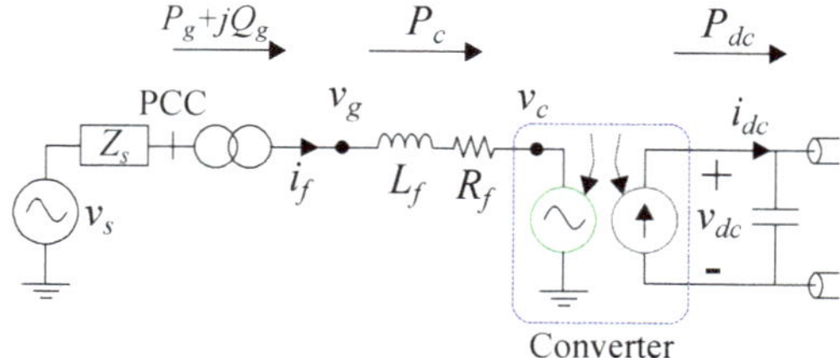

Figure 2. Average VSC model.

According to Figure 2 and considering the Kirchhoff's voltage law in the *abc* reference frame, the voltage drop along the impedance Z_s is:

$$v_g^{abc} - v_c^{abc} = L_f \frac{di_f^{abc}}{dt} + R_f i_f^{abc}. \tag{16}$$

Applying Park's Transformation $dq0$ and considering ω_g as the angular frequency of the rotating system; Equation (16) can be rewritten dividing the real and imaginary parts, giving the next expressions as a result [28]:

$$L_f \frac{di_f^d}{dt} = \omega_g L_f i_f^q - R_f i_f^d + v_g^d - v_c^d, \tag{17}$$

$$L_f \frac{di_f^q}{dt} = \omega_g L_f i_f^d - R_f i_f^q + v_g^q - v_c^q. \tag{18}$$

The d-axis is aligned to the $dq0$-rotating frame, which provokes the q-component equals zero in steady-state [29]. In this context, to synchronize the $dq0$-rotating frame a Phase-Locked Loop (PPL) must be used. Therefore, the real and reactive powers in the $dq0$-rotating frame can be expressed as follows:

$$P_g = v_g^d i_f^d \tag{19}$$

$$Q_g = -v_g^d i_f^q. \tag{20}$$

A VSC is capable of controlling independently active power P (or DC voltage) and reactive power Q (or AC voltage) [24]. Basically, the control consists of two control loops: the outer-control loop and the inner-control loop. Both are derived from the mathematical equations developed previously. The outer-control loop calculates the reference signals for i_f^{d*} and i_f^{q*}, which are employed as inputs to the inner-control loop. The inner-control loop gives the reference voltage [28] to V_c (controlled voltage source). Depending on the target of control, the VSC can work either as a rectifier or inverter. Considering a lossless converter, the power balance equation in the $dq0$ reference frame is [30]:

$$P_{dc} = P_{ac} = v_{dc} i_{dc} = v_g^d i_f^d + v_g^q i_f^q. \tag{21}$$

Expression (21) can also be rewritten as:

$$i_{dc} = \frac{v_g^d i_f^d + v_g^q i_f^q}{v_{cd}}. \tag{22}$$

In this case, Equation (22) provides the input signal to the controlled current source on the DC side of the VSC.

5.2. Model-Based Control

The model-based control of a VSC-based STATCOM model consists basically of two control loops. The first one corresponds to the outer loop depending on the variable to be controlled, V_{ac} or V_{dc}. The second one represents the inner loop based on the state-space model defined by expressions (17) and (18). Therefore, the conventional control uses a PI-controller, so that, the inner-control loop can be represented as:

$$v_c^d = \omega_g L_f i_f^q - \left(K_p - \frac{K_i}{s} \right) \left(i_f^{d*} - i_f^d \right) + v_g^d \tag{23}$$

$$v_c^q = -\omega_g L_f i_f^d - \left(K_p - \frac{K_i}{s} \right) \left(i_f^{q*} - i_f^q \right) + v_g^q. \tag{24}$$

The complete details of the conventional model-based control can be found in [31].

5.3. Data-Driven Controller

Even though data-driven controllers applied to VSC-based STATCOMs have been used in the past, most of them have been aimed at auxiliary control and not the converter itself [8,15]. Data-driven controllers enable a faster response which is an important feature demanded by any STATCOM model given that it needs to provide voltage support after an external event such as faults or connection/disconnection of loads.

The data-driven based STATCOM model proposed in this work is based on the response of the conventional VSC model. In fact, the data-driven controller consists only of inputs and outputs data and avoids the use of complex mathematical algorithms, which makes it suitable for a large power system that usually requires a high computational burden. The proposed data-driven strategy that is discussed in a general way in Section 4, is now described in terms of the VSC variables for its implementation.

To set-up the controller, it is not required to know any information about the models of the VSC and the grid. The only requirement is the definition of variables available for control and measurement, as well as a required operating point, that is, a set-point. We call these external variables accommodated in a vector w. To denote the desired value of the external variables at the equilibrium, that is, a set-point, we use the notation w^*. Then error variables are denoted by $\Delta w = w - w^*$. Then a set of measurements containing $w(1), w(2), \ldots, w(T)$ can be used to generate $\Delta w(1), \Delta w(2), \ldots, \Delta w(T)$, simply by subtracting the operating point value.

In this case, we selected Δi_d, Δi_q, Δv_{dc}, Δv_{ac}, and Δv_c^d, Δv_c^q as the external variable, since they are typically the available measurements in practice. Moreover, the variables Δv_c^d, Δv_c^q will permit implementation of the controller, as they provide the voltage reference for the VSC in terms of dq-components.

As described in Section 3.1, the collection of data on these variables permits us to obtain the coefficient matrix $\widetilde{P}$, which replaces the requirement of a model, since it contains sufficient information about the dynamics of the to-be-controlled system. Then, the matrix inequality shown in (11) is implemented using MATLAB and external optimization tools, in this case we used *Yalmip*. To set up (11), it is necessary to define the adequate sizes of $\widetilde{Psi}$ (in this case $N = 2$ and $q = 6$) and $\widetilde{V}$ (with the same dimension as $\widetilde{P}$. It thus remains to define the matrix $\widetilde{C}$ that contains the parameters of the candidate controller. For ease of implementation, we chose a PI configuration as the candidate controller with the following equations:

$$\Delta v_c^d = -K_1 \Delta i_d - K_2 \Delta i_q - K_3 \Delta v_{dc} - K_4 \Delta v_{ac} \\ - K_5 z_1 - K_6 z_2, \tag{25}$$

$$\Delta v_c^q = -G_1 \Delta i_d - G_2 \Delta i_q - G_3 \Delta v_{dc} - G_4 \Delta v_{ac} \\ - G_5 z_3 - G_6 z_4 , \tag{26}$$

where the variables $z_1, \ldots, z_4$ are obtained by discrete-time integration, that is,

$$\sigma \begin{bmatrix} z_1 \\ z_2 \\ z_3 \\ z_4 \end{bmatrix} = \begin{bmatrix} z_1 \\ z_2 \\ z_3 \\ z_4 \end{bmatrix} + \begin{bmatrix} \Delta v_{dc} \\ \Delta v_{ac} \\ \Delta v_{dc} \\ \Delta v_{ac} \end{bmatrix} . \tag{27}$$

The controller therefore has the following representation:

$$
\underbrace{\begin{bmatrix}
-K_1 & -K_2 & -K_3 & -K_4 & -1 & 0 & -K_5 & -K_6 & 0 & 0 \\
-G_1 & -G_2 & -G_3 & -G_4 & 0 & -1 & 0 & 0 & -G_5 & -G_6 \\
0 & 0 & -1 & 0 & 0 & 0 & \sigma-1 & 0 & 0 & 0 \\
0 & 0 & 0 & -1 & 0 & 0 & 0 & \sigma-1 & 0 & 0 \\
0 & 0 & -1 & 0 & 0 & 0 & 0 & 0 & \sigma-1 & 0 \\
0 & 0 & 0 & -1 & 0 & 0 & 0 & 0 & 0 & \sigma-1
\end{bmatrix}}_{C(\sigma)}
\begin{bmatrix}
\Delta i_d \\ \Delta i_q \\ \Delta v_{dc} \\ \Delta v_{ac} \\ \Delta v_c^d \\ \Delta v_c^q \\ z_1 \\ z_2 \\ z_3 \\ z_4
\end{bmatrix} = 0.
$$

As defined in (8), and as exemplified in (15), the factorization of $C(\sigma)$ leads to the coefficient matrix $\tilde{C}$. Finally, the gains $K_1, \ldots, K_6$ and $G_1, \ldots, G_6$ are numerically computed by solving (11). Although every version of the software is portable enough to work with different versions, in the present work, it might be of interest that we used MATLAB R2021a, YALMIP R20210331, and the standard LMILAB solver available in MATLAB. The latter was used as a default option, but the results were also corroborated by using SEDUMI 1.1 as a solver. The reader can refer to [32] for more information and more suitable options.

The realization of the controller in terms of a flow diagram is shown in Figure 3. In the following sections, we discuss the performance of the proposed data-driven controller with respect to the model-based strategy introduced in the previous section.

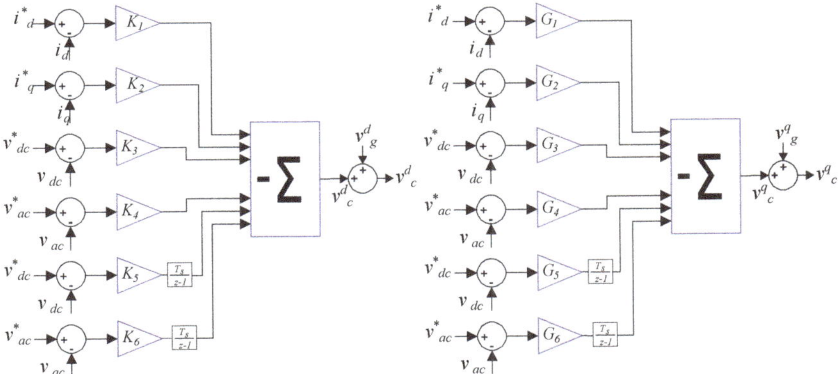

Figure 3. Data-driven controller for a VSC-based STATCOM model.

6. Case Study and Results

6.1. Test System

In order to validate the data-driven controller of a VSC-based STATCOM model, a two areas power system is employed. The test system is simulated in Matlab/Simulink software; it is comprised of a 2-area power system connected by two AC transmission lines and two machines in each area; these test systems are well-known, such as the Kundurs 2-area 4-machine power system. The complete details of the electrical grid can be found in [33]. The single-line diagram can be seen in Figure 4. The active power exchange between two areas is around 400 MW, going from area 1 to area 2, and the swing generator corresponds to machine 2, labeled as G2. The STATCOM model is connected to Bus 7 (B7) for controlling the AC bus voltage V_{ac} and the power ratings of the STATCOM can be seen in Appendix A, particularly in Table A1, where a step-up transformer 195 kV/230 kV is employed to connect it to the grid.

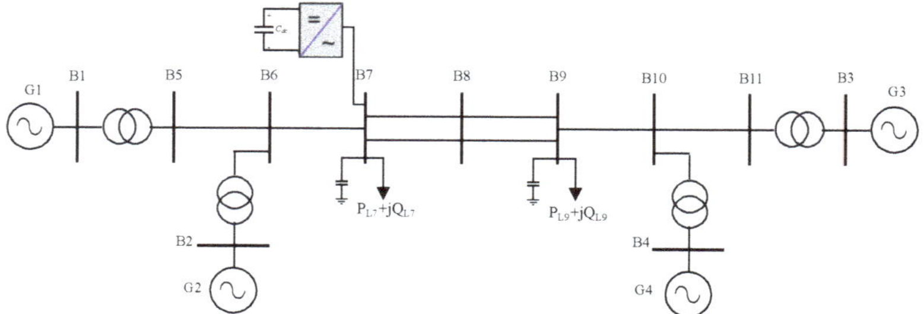

Figure 4. Two-area power system [33].

The proposed data-driven control is assessed considering different operating conditions according to the established voltage profile as well as transient faults to analyze power oscillations just after the fault clearing time. All simulations are carried out using Matlab/Simulink software, and a time step T_s equal to 1×10^{-4} is used for the implementation of the data-driven controller. The data-driven controller can be set up from the historical data collected from the grid or even from another model-based closed-loop operation of the STATCOM connected to the grid. Therefore, the data-driven controller will guarantee the dynamic performance of the STATCOM according to the rated reactive power because this will be mainly designed to operate to their limits.

The main advantage of the data-driven controller is that it is completely model-free, while its stability and general dynamic performance are equally deterministic, as if it was based on the existence of a completely accurate model of the grid. To account for this fact, we proceed to make a comprehensible comparison between the two scenarios: model-based and model-free. Other advantages include the ability to bypass the issue arising from model uncertainties, which is the typical weakness of any model-based approach. Based on the described advantages, the proposed approach is validated analyzing the performance of a VSC-based STATCOM model, which is compared to the model-based control system under three different scenarios: (a) by using changes in AC reference voltage; (b) by analyzing the voltage recovery after a three-phase fault; (c) power oscillation damping after the fault clearing time. For either, data-driven and conventional models, an average model of the VSC is used. The PPL parameters and grid parameters are kept the same for both types of controllers.

6.2. Voltage Step Response

To demonstrate the performance of the proposed method, we first use a test for comparing both model-based and data-driven control, which consists of changing the AC voltage reference V_{ac}. For this scenario, V_{dc}^* remains unchanged given that it should be constant according to [34]; this means that I_d and I_q should be effectively compensated for by the controller in order to keep V_{ac} and V_{dc} close to their references.

Figure 5a shows the results for different set points of AC voltages. First, the V_{ac} is stepped up from 1.006 to 1.044 p.u. at $t = 2$ s; then, it is also stepped down from 1.044 to 0.98 p.u. at $t = 8$ s; finally, at $t = 12$ s the reference changes from 0.97 to 1.022 p.u. Notice that both controls present similar behavior during the changes of voltage; this can be confirmed with the error of differences shown in Figure 5b. Faster response time represents a faster time to reach a new steady-state defined by a particular variable due to reference changes, which can be measured with Δt. Figure 6 shows a fair comparison of the model-based controller and the proposed method during the first voltage step shown in Figure 5. Notice that Δt_1 corresponds to the time in which the data-driven controller reaches the new steady-state, while Δt_2 is the elapsed time by the conventional model-based control. In conclusion, Δt_1 corresponds to the settling time which is around 0.242 s, whilst Δt_2 is close to 0.426. Based on the results, the VSC-based STATCOM model with a conventional controller takes more time to reach the new steady-state compared to the data-driven controller. Other time steps are also used to analyze the dynamic response of the controller, and these are summarized in Table 1.

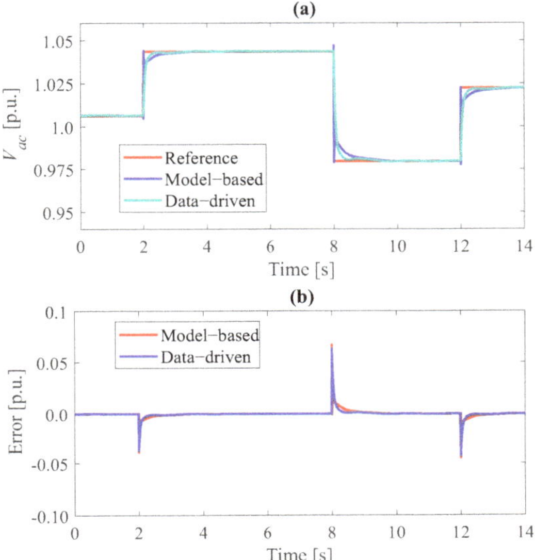

Figure 5. Reference changes: (**a**) voltage step comparison and (**b**) errors of differences.

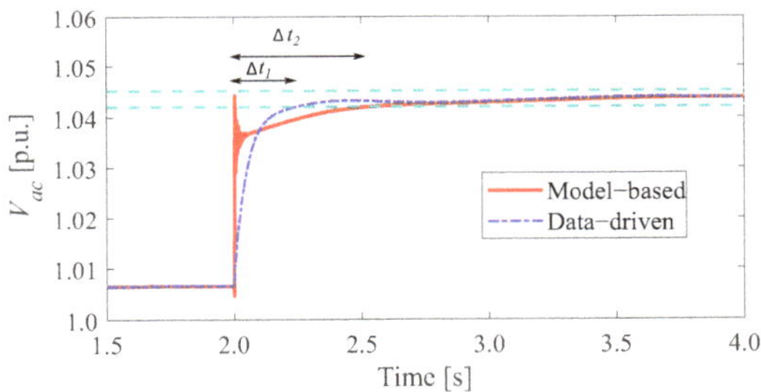

Figure 6. Comparison of both control approaches during voltage reference changes.

Table 1. Controller performance using different time steps.

Step Time	Settling Time	
	Model-Based	**Data-Driven**
$0.25T_s$	0.528	0.243
$0.5T_s$	0.527	0.242
T_s	0.426	0.242
$2T_s$	0.425	0.242

Figure 7 displays the current flowing from the STATCOM to the grid, current components in the dq reference frame. Notice that a change in the voltage reference does not impact the current component i_d significantly as shown in Figure 7a, while the second component i_q presents the most noticeable changes due to a voltage change may demand more reactive power, and this will be reflected in the reactive component of the current flowing to the system. Figure 8 presents the dynamical performance of the reactive power, which is injected into the grid according to the objective of control, that is, a higher voltage will be demanding more reactive power (capacitive) and vice-versa.

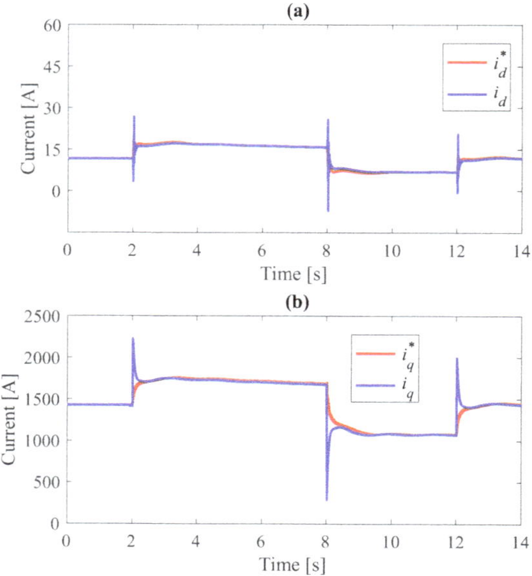

Figure 7. Currents of the data-driven controller during changes in the AC reference voltage: (**a**) *d*-axis current and (**b**) *q*-axis current.

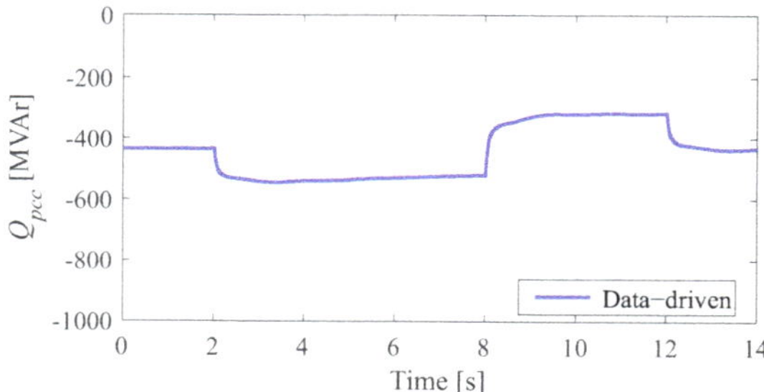

Figure 8. Reactive power at PCC using the data-driven controller with changes in the AC reference voltage.

6.3. Voltage Recovery under Transient Faults

Another subject of interest to assess the dynamic performance of a STATCOM is linked to its ability to provide a fast voltage recovery after a transient event by injecting reactive power, which results in improving the power system stability limits [35,36]. In this context, the VSC-based STATCOM model using a data-driven controller is analyzed because the lack of reactive power may deteriorate the bus voltage values as well the power transfer limits [37]. In this work, the voltage recovery is assessed under two different scenarios: (1) a solid-grounded three-phase fault on Bus 7, and (b) a solid-grounded three-phase fault in one of the parallel transmission lines. For both scenarios, the clearing time corresponds to 100 ms.

6.3.1. Transient Fault on Bus 7

Figure 9 shows the dynamic response during a transient fault simulated at Bus 7. Notice that both controllers present small differences in the voltage control. However, a better performance can be observed when the data-driven controller is employed. From Figure 9, one can notice that the proposed control presents a higher voltage overshoot than the model-based control; however, once the fault is cleaned up, the data-driven controller reaches the steady-state faster than the conventional controller. In addition, during any transient fault, the STATCOM will act to maintain the voltage profile according to the AC voltage reference and depending on the fault severity, where sometimes the controller can be saturated. This will be defined by the reactive power requirements during the pre-fault condition, where a larger AC voltage reference will be demanding more reactive power. A fault may also produce a higher overshoot during the transient response, and the controller may be saturated because this has physical limits.

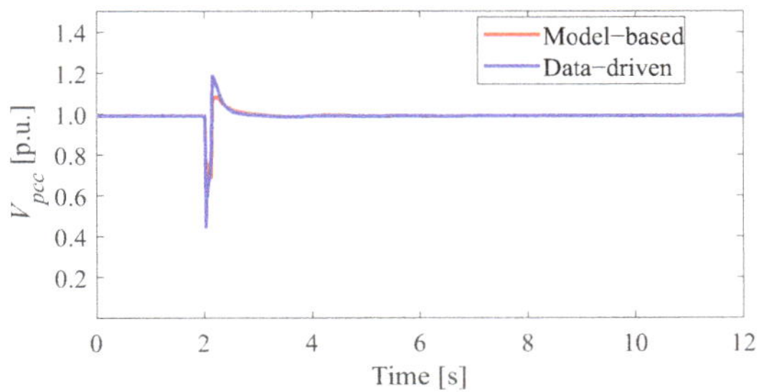

Figure 9. Voltage control during transient fault at Bus 7.

Figure 10 confirms the robustness of the proposed approach because the power flows between the interconnected areas, active and reactive powers, present smaller power oscillations than the generated with the conventional control. For instance, Figure 10a shows significant differences during the transient behavior of the active power, while Figure 10b displays the results of the reactive power.

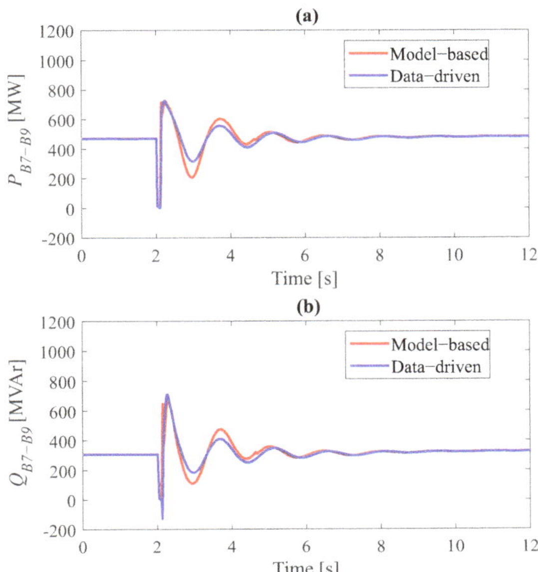

Figure 10. Power flow during a fault at Bus 7: (**a**) active power, and (**b**) reactive power.

On the other hand, a STATCOM not only offers the capability to improve the power system efficiency due to its fast dynamic response of voltage control but can also help to mitigate low-frequency power oscillations [38]. For example, Figure 11 displays the difference between two rotor angles corresponding to Generator 1 and Generator 2 (defined as slack generator). The results show low-frequency oscillations between both generators when a PSS is used for every generator, except for Generator 4. For all analyzed scenarios, the machine speed deviation is used as an input signal to each PSS. According to the results shown in Figure 11, notice that the proposed controller helps to reduce the power oscillations due to the fast dynamic response to recover the AC voltage at the PCC. In addition, the differences shown in Figure 11 not only depend on the STATCOM but are also due to other generators. This is the main reason that both responses do not match very well during the transient period. However, notice that after some seconds both controllers have the same behavior; this means that the power system has reached the new steady-state.

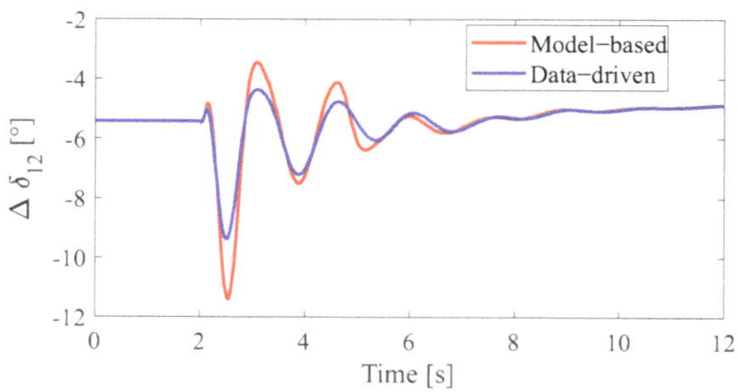

Figure 11. Rotor angles during a fault at Bus 7.

6.3.2. Fault along the Transmission Line

In this case, a transient fault is analyzed, which is cleaned up by opening the circuit-breakers of the faulted transmission line. The dynamic response corresponding to voltage at the PCC is depicted in Figure 12, where significant differences occur between the proposed control scheme and the model-based control. In addition, Figure 13 shows the power flow changes after the fault clearing time. The power flow measurements are taken from the non-faulted transmission line. The power transmission losses are increased due to the presence of only one transmission line. The damping capability is highly noticeable in Figure 14, which shows the rotor angle difference after the clearing time and due to the change of topology caused by the opening of one transmission line. Figures 11 and 14 help to confirm the dynamic performance of the data-driven controller in comparison with the conventional model-based controller. Notice that a better performance is exhibited when the STATCOM is controlled by the data-driven approach. Finally, the described results help to confirm the dynamic performance of the data-driven controller in comparison with the model-based controller, where significant differences appear during the transient period. Table 2 summarizes the controller performance during transient faults after evaluating different time steps; voltage recovery after the fault clearing time. Finally, considering all analyzed variables, the results showed that the data-driven controller offers better performance during transient events because the resulting power oscillations are smaller in magnitude for all analyzed scenarios.

Table 2. Controller performance during transient faults using different time steps.

Step Time	Overshoot		Settling Time	
	Model-Based	Data-Driven	Model-Based	Data-Driven
$0.1T_s$	0.100	0.21	1.050	0.84
$0.5T_s$	0.093	0.21	1.030	0.78
T_s	0.090	0.21	0.894	0.83
$2T_s$	0.090	0.20	0.894	0.83

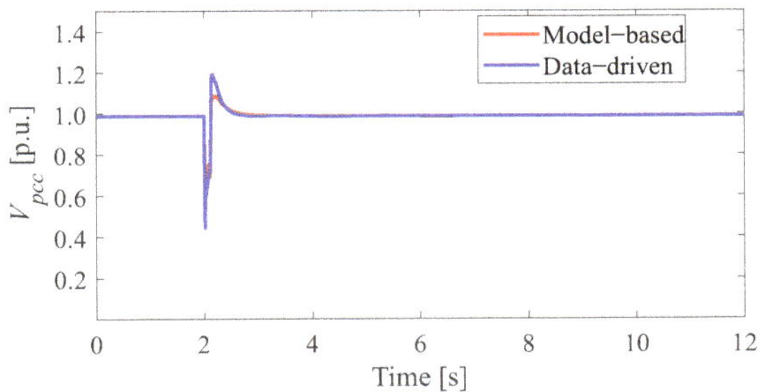

Figure 12. AC voltage during a fault on the transmission line.

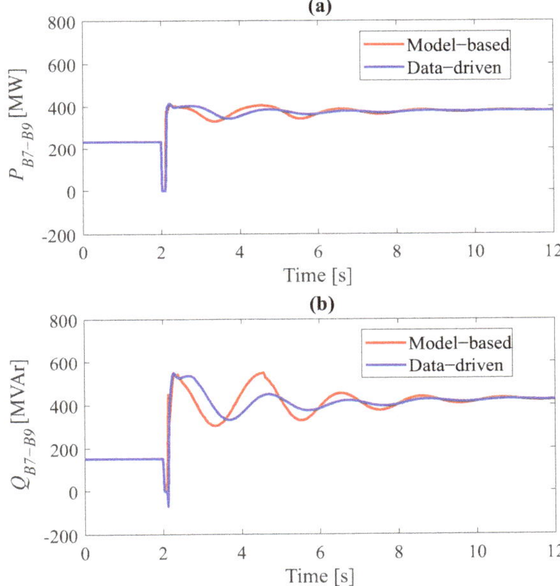

Figure 13. Power flows during transient fault and topology change: (**a**) active power, and (**b**) reactive power.

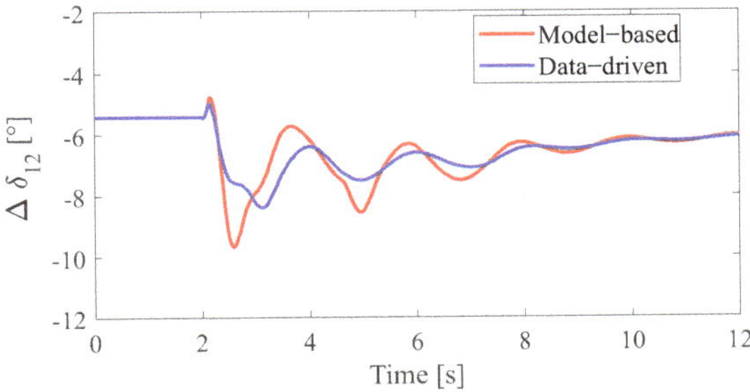

Figure 14. Rotor angles during transient fault and topology change.

6.4. Load Shedding Assessment

One of the last resources to mitigate electric power generation deficiency is load shedding. It consists of disconnecting the less essential loads connected to the grid. An inequality between power generation and consumption affects the power system frequency leading to a collapse [39]. Load shedding is a common practice that can be either beneficial or detrimental to the power system stability. The disconnection of considerable sizing loads creates a mismatch between mechanical and electrical power, causing a positive power acceleration that can lead to power system instability.

To assess the performance of the STATCOM after a load shedding, a 200 MW load is disconnected from Bus 9. The load shedding decreases the power flow between the two areas bringing a new condition to the generation (rotor angles). Figures 15 and 16 show the comparison between the transient response of the model-based and data-driven controllers. Regarding the real and reactive power flows shown in Figure 15, both controllers present quite a similar performance. The transient response of the AC voltage shown in Figure 16 reaches a peak voltage of 1.03 p.u. when the data-driven controller is employed. That voltage is a bit higher than that produced by the model-based controller but both controllers match very well due to both producing the same magnitude on the first oscillation and almost the same settling time. The reactive power injected by the STATCOM can be shown in Figure 16, where both controllers present a similar behavior. Notice that the load shedding will demand less reactive power as expected.

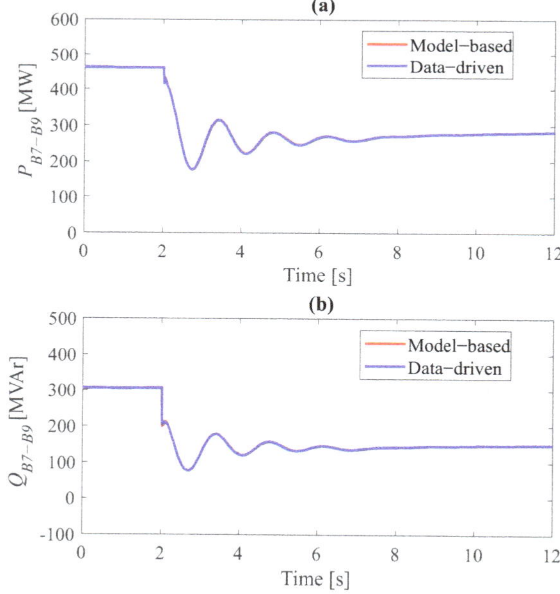

Figure 15. Power flows during load shedding: (**a**) active power, and (**b**) reactive power.

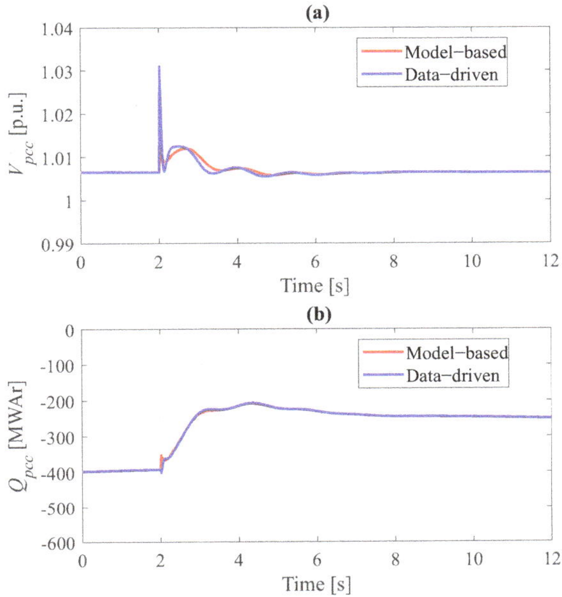

Figure 16. STATCOM: (**a**) AC voltage, and (**b**) reactive power during load shedding.

7. Conclusions

A new control approach based on data-driven was proposed. The fundamentals of design were included, aiming to develop a new control strategy for a STATCOM model. The proposed data-driven control was assessed, considering different operating conditions according to the established voltage profile as well as transient faults to analyze power oscillations just after the fault clearing time. A solid comparison of a VSC-based STATCOM model was developed between a model-based control and a data-driven control approach. The studies were focused on voltage control and power oscillation damping capabilities. For voltage control, two tests were carried out: (a) voltage step response, and (b) voltage recovery under fault scenarios. After testing the scenarios, the authors concluded that the data-driven controller showed a better performance in all scenarios compared to the conventional model-based controller, providing a faster response for voltage control and power oscillations damping.

Author Contributions: Conceptualization, D.R. and D.G.; methodology, D.R., D.G. and J.C.M.-M.; validation, D.R., D.G. and J.C.M.-M.; formal analysis, D.R., J.E.V.-R. and G.E.; investigation, D.R.; data curation, J.C.M.-M. and J.E.V.-R.; writing—original draft preparation, D.R., D.G., J.C.M.-M., J.E.V.-R. and G.E.; writing—review and editing, J.E.V.-R. and G.E.; supervision, D.G. and J.C.M.-M. All authors have read and agreed to the published version of the manuscript.

Funding: This research is a product of the Project 266632 "Laboratorio Binacional para la Gestión Inteligente de la Sustentabilidad Energética y la Formación Tecnológica" ["Bi-National Laboratory on Smart Sustainable Energy Management and Technology Training"], funded by the CONACYT-SENER Fund for Energy Sustainability (Agreement: S0019-2014-01).

Conflicts of Interest: The authors declare no conflict of interest.

Abbreviations

The following abbreviations are used in this manuscript:

RES	Renewable Energy Resources
FACTS	Flexible AC Transmission System
STATCOM	Static Synchronous Compensator
POD	Power Oscillation Damping
PSS	Power System Stabilizer
PCC	Point of Common Coupling
PV	Photovoltaic
ANFIS	Adaptive Network-Based Fuzzy Inference System
WNN	Wavelet Neural Network
ADC	Additional damper controller
DDPG	Deep Deterministic Policy Gradient
VSC	Voltage Source Converter
PLL	Phase-Locked Loop
DC	Direct Current
AC	Alternating Current
LMIs	Linear Matrix Inequalities

Appendix A

The STATCOM parameters and ratings are a modified version from the one used in [24,40], which are presented in Table A1.

Table A1. Parameters of the VSC-based STATCOM.

Parameters	Value
Rated power	550 MVA
Rated alternating voltage	195 kV
Rated direct voltage	$\pm$150 kV
Coupling resistance	1.0864 Ω
Coupling inductance	0.0692 H
DC capacitor per pole	114 μF
Converter transformer ratio	230 kV/195 kV

References

1. Li, W.; Wang, Y.; Chen, T. Investigation on the thevenin equivalent parameters for online estimation of maximum power transfer limits. *IET Gener. Transm. Distrib.* **2010**, *4*, 1180–1187. [CrossRef]
2. Castro, L.M.; Acha, E.; Fuerte-Esquivel, C.R. A Novel STATCOM Model for Dynamic Power System Simulations. *IEEE Trans. Power Syst.* **2013**, *28*, 3145–3154. [CrossRef]
3. Varma, R.K.; Salehi, R. SSR Mitigation With a New Control of PV Solar Farm as STATCOM (PV-STATCOM). *IEEE Trans. Sustain. Energy* **2017**, *8*, 1473–1483. [CrossRef]
4. Ahmadinia, M.; Ghazi, R. Coordinated Control of STATCOM and ULTC to Reduce Capacity of STATCOM. In Proceedings of the Iranian Conference on Electrical Engineering (ICEE), Mashhad, Iran, 8–10 May 2018; pp. 1062–1066. [CrossRef]
5. Ayala-Chauvin, M.; Kavrakov, B.S.; Buele, J.; Varela-Aldás, J. Static Reactive Power Compensator Design, Based on Three-Phase Voltage Converter. *Energies* **2021**, *14*, 2198. [CrossRef]
6. Fan, R.; Wang, S.; Huang, R.; Lian, J.; Huang, Z. Wide-area measurement-based modal decoupling for power system oscillation damping. *Electr. Power Syst. Res.* **2020**, *178*, 106022. [CrossRef]
7. AL-Ismail, F.S.; Hassan, M.A.; Abido, M.A. RTDS Implementation of STATCOM-Based Power System Stabilizers. *Can. J. Electr. Comput. Eng.* **2014**, *37*, 48–56. [CrossRef]
8. Zhang, G.; Hu, W.; Cao, D.; Yi, J.; Huang, Q.; Liu, Z.; Chen, Z.; Blaabjerg, F. A data-driven approach for designing STATCOM additional damping controller for wind farms. *Int. J. Electr. Power Energy Syst.* **2020**, *117*, 105620. [CrossRef]
9. Liu, J.; Xu, Y.; Dong, Z.Y.; Wong, K.P. Retirement-Driven Dynamic VAR Planning for Voltage Stability Enhancement of Power Systems With High-Level Wind Power. *IEEE Trans. Power Syst.* **2018**, *33*, 2282–2291. [CrossRef]
10. Varma, R.K.; Maleki, H. PV Solar System Control as STATCOM (PV-STATCOM) for Power Oscillation Damping. *IEEE Trans. Sustain. Energy* **2019**, *10*, 1793–1803. [CrossRef]
11. Kumar, V.; Pandey, A.S.; Sinha, S.K. Stability Improvement of DFIG-Based Wind Farm Integrated Power System Using ANFIS Controlled STATCOM. *Energies* **2020**, *13*, 4707. [CrossRef]

12. Beza, M.; Bongiorno, M. An Adaptive Power Oscillation Damping Controller by STATCOM With Energy Storage. *IEEE Trans. Power Syst.* **2015**, *30*, 484–493. [CrossRef]
13. Dilshad, S.; Abas, N.; Farooq, H.; Kalair, A.R.; Memon, A.A. NeuroFuzzy Wavelet Based Auxiliary Damping Controls for STATCOM. *IEEE Access* **2020**, *8*, 200367–200382. [CrossRef]
14. Peres, W.; da Costa, N.N. Comparing strategies to damp electromechanical oscillations through STATCOM with multi-band controller. *ISA Trans.* **2020**, *107*, 256–269. [CrossRef]
15. Noh, H.; Cho, H.; Lee, S.; Lee, B. STATCOM with SSR damping controller using geometric extraction on phase space reconstruction method. *Int. J. Electr. Power Energy Syst.* **2020**, *120*, 106017. [CrossRef]
16. Liu, Y.; Huang, A.Q.; Song, W.; Bhattacharya, S.; Tan, G. Small-signal model-based control strategy for balancing individual DC capacitor voltages in cascade multilevel inverter-based STATCOM. *IEEE Trans. Ind. Electron.* **2009**, *56*, 2259–2269. [CrossRef]
17. Xu, Y.; Li, F. Adaptive PI control of STATCOM for voltage regulation. *IEEE Trans. Power Deliv.* **2014**, *29*, 1002–1011. [CrossRef]
18. Emam, A.S.; Azmy, A.M.; Rashad, E.M. Enhanced Model Predictive Control-Based STATCOM Implementation for Mitigation of Unbalance in Line Voltages. *IEEE Access* **2020**, *8*, 225995–226007. [CrossRef]
19. Zhang, Y.; Yuan, X.; Wu, X.; Yuan, Y.; Zhou, J. Parallel Implementation of Model Predictive Control for Multilevel Cascaded H-Bridge STATCOM With Linear Complexity. *IEEE Trans. Ind. Electron.* **2020**, *67*, 832–841. [CrossRef]
20. Hashemzadeh, E.; Khederzadeh, M.; Aghamohammadi, M.R.; Asadi, M. A Robust Control for D-STATCOM Under Variations of DC-Link Capacitance. *IEEE Trans. Power Electron.* **2021**, *36*, 8325–8333. [CrossRef]
21. Yang, S.; Lei, Q.; Peng, F.Z.; Qian, Z. A Robust Control Scheme for Grid-Connected Voltage-Source Inverters. *IEEE Trans. Ind. Electron.* **2011**, *58*, 202–212. [CrossRef]
22. Gardner, R.F. *Introduction to Plant Automation and Controls*; CRC Press: Boca Raton, FL, USA, 2020.
23. Cominos, P.; Munro, N. PID controllers: Recent tuning methods and design to specification. *IEE Proc.-Control Theory Appl.* **2002**, *149*, 46–53. [CrossRef]
24. Wu, G.; Liang, J.; Zhou, X.; Li, Y.; Egea-Alvarez, A.; Li, G.; Peng, H.; Zhang, X. Analysis and design of vector control for VSC-HVDC connected to weak grids. *CSEE J. Power Energy Syst.* **2017**, *3*, 115–124. [CrossRef]
25. Willems, J.C.; Polderman, J.W. *Introduction to Mathematical Systems Theory: A Behavioral Approach*; Springer Science & Business Media: Berlin/Heidelberg, Germany, 1997; Volume 26.
26. Willems, J.C.; Rapisarda, P.; Markovsky, I.; De Moor, B.L. A note on persistency of excitation. *Syst. Control Lett.* **2005**, *54*, 325–329. [CrossRef]
27. Imhof, M. Voltage Source Converter Based HVDC—Modelling and Coordinated Control to Enhance Power System Stability. Ph.D. Thesis, ETH Zurich, Zurich, Switzerland, 2015.
28. Kumaravel, S.; Narayan, R.S.; O'Donnell, T.; O'Loughlin, C. Genetic algorithm based PI tuning of VSC-HVDC system and implementation using OPAL-RT. In Proceedings of the TENCON 2017—2017 IEEE Region 10 Conference, Penang, Malaysia, 5–8 November 2017; pp. 2193–2197. [CrossRef]
29. Imhof, M.; Andersson, G. Dynamic modeling of a VSC-HVDC converter. In Proceedings of the 48th International Universities' Power Engineering Conference (UPEC), Dublin, Ireland, 2–5 September 2013; pp. 1–6. [CrossRef]
30. Jovcic Khaled Ahmed, D. Two-level VSC HVDC Modelling, Control, and Dynamics. In *High Voltage Direct Current Transmission*; John Wiley & Sons, Ltd.: Hoboken, NJ, USA, 2019; Chapter 17, pp. 227–245. [CrossRef]
31. Li, C.; Burgos, R.; Wen, B.; Tang, Y.; Boroyevich, D. Stability Analysis of Power Systems with Multiple STATCOMs in Close Proximity. *IEEE Trans. Power Electron.* **2020**, *35*, 2268–2283. [CrossRef]
32. Löfberg, J. YALMIP: A Toolbox for Modeling and Optimization in MATLAB. In Proceedings of the CACSD Conference, Taipei, Taiwan, 2–4 September 2004.
33. Kundur, P. *Power System Stability and Control*; McGraw Hill Education: New York, NY, USA, 1994.
34. Ramirez, D.; Herrero, L.C.; de Pablo, S.; Martinez, F. STATCOM Control Strategies. In *Static Compensators (STATCOMs) in Power Systems*; Shahnia, F., Rajakaruna, S., Ghosh, A., Eds.; Springer: Singapore, 2015; pp. 147–186. [CrossRef]
35. Bhole, S.; Nigam, P. Improvement of Voltage Stability in Power System by Using SVC and STATCOM. *Int. J. Adv. Res. Electr. Electron. Instrum. Eng.* **2015**, *4*, 749–755. [CrossRef]
36. Neutz, M. Power Quality. In *Voltage Stabilisation for Industrial Grids and Wind Farms with STATCOM*; ABB: Zurich, Switzerland, 2013; p. 38.
37. Barua, P.; Quamruzzaman, M. Steady State Voltage Vulnerability and Stability Limit Analysis of Bangladesh Power System Using STATCOM as a Shunt Compensator. In Proceedings of the 4th International Conference on Electrical Engineering and Information Communication Technology (iCEEiCT), Dhaka, Bangladesh, 13–15 September 2018; pp. 1–4. [CrossRef]
38. Huang, W.; Sun, K. Optimization of SVC settings to improve post-fault voltage recovery and angular stability. *J. Mod. Power Syst. Clean Energy* **2019**, *7*, 491–499. [CrossRef]
39. Abou, A.A.; El-Din, A.Z.; Spea, S.R. Optimal load shedding in power systems. In Proceedings of the Eleventh International Middle East Power Systems Conference, El-Minia, Egypt, 19–21 December 2006; Volume 2, pp. 568–575.
40. Harnefors, L.; Nee, H.P. Modelling and control of VSC-HVDC connected to island systems. *IEEE Trans. Power Syst.* **2010**, 783–793. [CrossRef]

MDPI

Article

Adaptive Robust Motion Control of Quadrotor Systems Using Artificial Neural Networks and Particle Swarm Optimization

Hugo Yañez-Badillo [1,*], Francisco Beltran-Carbajal [2], Ruben Tapia-Olvera [3], Antonio Favela-Contreras [4], Carlos Sotelo [4] and David Sotelo [4]

[1] Departamento de Investigación, Tecnológico de Estudios Superiores de Tianguistenco, Santiago Tilapa 52650, Mexico

[2] Departamento de Energía, Universidad Autónoma Metropolitana, Unidad Azcapotzalco, Mexico City 02200, Mexico; fbeltran@azc.uam.mx

[3] Departamento de Energía Eléctrica, Universidad Nacional Autónoma de México, Mexico City 04510, Mexico; rtapia@fi-b.unam.mx

[4] Tecnologico de Monterrey, School of Engineering and Science, Ave. Eugenio Garza Sada 2501, Monterrey 64849, Mexico; antonio.favela@tec.mx (A.F.-C.); carlos.sotelo@tec.mx (C.S.); david.sotelo@tec.mx (D.S.)

* Correspondence: hugo_mecatronica@test.edu.mx

Citation: Yañez-Badillo, H.; Beltran-Carbajal, F.; Tapia-Olvera, R.; Favela-Contreras, A.; Sotelo, C.; Sotelo, D. Adaptive Robust Motion Control of Quadrotor Systems Using Artificial Neural Networks and Particle Swarm Optimization. *Mathematics* **2021**, *9*, 2367. https://doi.org/10.3390/math9192367

Academic Editor: Alfonso Baños

Received: 3 August 2021
Accepted: 10 September 2021
Published: 24 September 2021

Publisher's Note: MDPI stays neutral with regard to jurisdictional claims in published maps and institutional affiliations.

Abstract: Most of the mechanical dynamic systems are subjected to parametric uncertainty, unmodeled dynamics, and undesired external vibrating disturbances while are motion controlled. In this regard, new adaptive and robust, advanced control theories have been developed to efficiently regulate the motion trajectories of these dynamic systems while dealing with several kinds of variable disturbances. In this work, a novel adaptive robust neural control design approach for efficient motion trajectory tracking control tasks for a considerably disturbed non-linear under-actuated quadrotor system is introduced. Self-adaptive disturbance signal modeling based on Taylor-series expansions to handle dynamic uncertainty is adopted. Dynamic compensators of planned motion tracking errors are then used for designing a baseline controller with adaptive capabilities provided by three layers B-spline artificial neural networks (Bs-ANN). In the presented adaptive robust control scheme, measurements of position signals are only required. Moreover, real-time accurate estimation of time-varying disturbances and time derivatives of error signals are unnecessary. Integral reconstructors of velocity error signals are properly integrated in the output error signal feedback control scheme. In addition, the appropriate combination of several mathematical tools, such as particle swarm optimization (PSO), Bézier polynomials, artificial neural networks, and Taylor-series expansions, are advantageously exploited in the proposed control design perspective. In this fashion, the present contribution introduces a new adaptive desired motion tracking control solution based on B-spline neural networks, along with dynamic tracking error compensators for quadrotor non-linear systems. Several numeric experiments were performed to assess and highlight the effectiveness of the adaptive robust motion tracking control for a quadrotor unmanned aerial vehicle while subjected to undesired vibrating disturbances. Experiments include important scenarios that commonly face the quadrotors as path and trajectory tracking, take-off and landing, variations of the quadrotor nominal mass and basic navigation. Obtained results evidence a satisfactory quadrotor motion control while acceptable attenuation levels of vibrating disturbances are exhibited.

Keywords: quadrotor UAV; artificial neural networks; robust control; Taylor series; B-splines; particle swarm optimization

1. Introduction

It is known that, in motion control systems, it is required that the system move to match some desired features of acceleration, velocity, position, or a combination of them. Unmanned aerial vehicles (UAVs) are dynamic systems where the controlled motion is fundamental to complete specific applications. Recently, diverse types of UAVs vehicles have been developed, with fixed-wing unmanned aerial vehicles (FW-UAVs) being the most common and most developed. These aircraft are similar to passenger aircraft, with a pair of wings to provide lift, a propellant system to provide thrust, and aerodynamic surfaces to control the motion. Their efficiency is higher compared to other UAVs, allowing it to perform long flights. Nevertheless, their indoors use is exclude since they do not have the ability to hovering and can not turn at reduced distances [1]. For their part, rotary-wing unmanned aerial vehicles (RW-UAVs) have various configurations including the conventional helicopter, the coaxial helicopter, and multi-rotors, which can sustain hover flight and take-off-landing vertically (VTOL). The FW-UAVs and RW-UAVs are the classic configurations most used in the applications assigned to unmanned aerial vehicles. Among the main ones are surveillance, monitoring, photography, inspection, and cargo transportation [2], with RW-UAVs having more civil applications than FW-UAVs [3]. On the other hand, technological advances have also allowed the development of new UAV configurations, such as bio-inspired flapping-wing unmanned aerial vehicles (Fl-UAV) [4] and lighter-than-air unmanned aerial vehicles, (LtA-UAVs) [5]. The four rotor helicopter or quadrotor is the most common rotorcraft platform in the research community due to its properties of under-actuation, low construction cost, symmetrical structure, high coupling non-linear dynamics, and capabilities of VTOL and hovering.

In the literature, several important contributions have been reported for controlling the quadrotor dynamics. Motion controllers based on theories, such as sliding modes [6], active disturbance rejection [7], backstepping [8], Lyapunov functions [9], H_∞ [10], adaptive controllers based on $\mathcal{L}_1$ [11,12], fuzzy logic [13], neural networks [14], model predictive control [15], or combination of them. Since, to some, drawbacks are inherent to each control strategy, such as high-frequency control actions, unmeasurable system information required, high dependency of mathematical models, high-gain feedback, and high sensibility against exogenous disturbances, some researchers have been properly exploited the properties of adaptive and robust control for designing advanced control methodologies.

In contrast with conventional control, intelligent control techniques are able to efficiently deal with incomplete information of many dynamic systems and its environment within a wide range of operational conditions. Then, adaptive control strategies represent a potential alternative for improving the performance of robust motion control schemes. In the literature, adaptive control stands for a class of control techniques used for compensating parameter changes, disturbances, and unknown changes in the system, by adaptations based on observations [16]. Relevant and recently research have been inspired by the qualities of adaptive and robust control schemes for quadrotor motion control. Authors in [17] introduce a model reference adaptive control scheme for a four-rotor helicopter in order to increase robustness against parametric uncertainty. A baseline controller is proposed for trajectory tracking task which is further improved by including adaptive capabilities. Similarly, switched adaptive controller are properly introduced in [18,19]. Here, controllers are suitably designed for controlling a quadrotor in the presence of unknown external disturbances and variations in the mass and inertia of the quadrotor due to unknown payload. Strict simulation scenarios are brought out to validate their proposal.

On the other hand, another adaptive control scheme is presented by authors in [20], where the quadrotor attitude is stabilized by an adaptive multi-variable finite-time algorithm. The controller design is carried out by using an improved super-twisting technique. Control methodologies designed based on the central ideas of adaptive sliding mode control are presented in [21,22]. In [21], a disturbance observer (DO) is integrated in control design to compensate external disturbances. The tune of the gain of sliding surface is accomplished via neural networks. In contrast, authors in [22] implement an adaptive scheme by proposing a super twisting controller along with Lyapunov-based function methodology and discontinuous projection operators. The research in [23] presents a fuzzy adaptive linear active disturbance rejection controller. The fuzzy framework is setup successfully to compute the observer bandwidth, controller bandwidth, as well as the control compensation factor.

Considering the aforementioned information, in this paper, authors introduce a novel and efficient adaptive robust motion tracking control for quadrotor non-linear systems. The main differences with others proposals reported in the literature are enlisted below:

1. Only position measurements are required for feedback control;
2. High-gain feedback is reduced by using B-spline artificial neural networks;
3. Reduced amount of control parameters needs to be tuned;
4. The use of disturbance observers is unnecessary;
5. The use of the tracking error derivatives is avoided in the controller design;
6. Offline training of B-spline artificial neural networks is performed by particle swarm optimization;
7. Low dependency of the quadrotor non-linear mathematical model;
8. Robustness against a class of external disturbances, including undesirable vibrating forces and torques.

The content of this paper is summarized as follows: the quadrotor non-linear and high coupling mathematical model is presented in Section 2. In Section 3, the design procedure of the novel robust and adaptive controller is introduced. Subsequently, some simulation experiments are presented in Section 4 in order to highlight the performance of the introduced methodology. Finally, some conclusions, remarks, and future work are mentioned in conclusions section.

2. Mathematical Quadrotor Model

A quadrotor is an aerial under-actuated mechanical system with four independent variable speed rotors. It has six degrees of freedom which are controlled by four control inputs: a main thrust force (u), and three torques (rolling τ_ϕ, pitching τ_θ, and yawing τ_ψ). Lateral, longitudinal, and vertical motion are achieved by a suitable combination of the control inputs. The main force, produced by the total sum of the thrust provided by each individual rotor, allows the quadrotor to take-off and land, as well as hover. Meanwhile, control torques are generated when there exists a difference of the produced forces by two pair of rotors: first pair rotating clockwise is formed by rotors 1 and 3, and the second by 2 and 4 rotors spinning in the opposite direction. Different from conventional helicopters it is not required a mechanical pitch system for the rotor blades.

The Euler–Lagrange and Newton–Euler formalisms are usually used to obtain the quadrotor dynamics described by a set of highly coupled non-linear differential equations. The quadrotor pose is determined by considering a body-fixed frame with X', Y', and Z' axes coincident with the centre of mass, and a global inertial coordinate system, or earth-fixed frame, with X, Y, and Z axes, as shown in Figure 1. By nature the quadrotor is an unstable system, and during outdoor and indoor flying, quadrotors might be subjected to undesirable vibrating disturbances. Thus, it should be designed efficient force and torque controllers to perform a proper motion tracking in the three-dimensional space.

The control inputs are related with each individual rotor by the following expressions

$$u = \sum_{i=1}^{4} F_i$$

$$\tau_\psi = \sum_{i=1}^{4} \tau_{M_i}$$

$$\tau_\theta = (F_3 - F_1)l$$

$$\tau_\phi = (F_2 - F_4)l \tag{1}$$

here l stands for the distance measured from a rotor axis to the quadrotor centre of mass, and τ_{M_i} is the couple developed by motor M_i. F_i and τ_{M_i} are functions of the rotor angular velocities. From Equation (1) it is appreciated that the quadrotor motion is possible by suitably combining the control inputs.

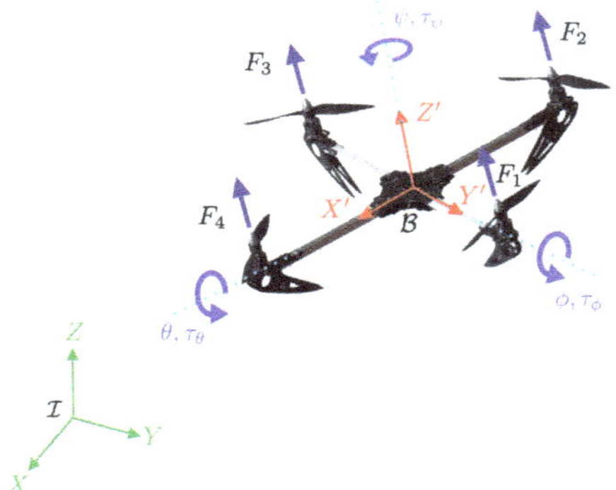

Figure 1. Schematic of a non-linear quadrotor system [24].

In this work, the non-linear dynamic model of the quadrotor is derived by means of the Euler–Lagrange formalism. In order to describe the system, let us consider the following vector of generalized coordinates

$$\mathbf{q} = [x\ y\ z\ \phi\ \theta\ \psi]^{\top} \in \mathbb{R}^6 \tag{2}$$

the centre of mass position is represented by the variables x, y, and z, and the quadrotor attitude is described by the set of Euler roll ϕ, pitch θ, and yaw ψ angles.

The Lagrangian is defined by the difference of the kinetic and potential energy, so we get

$$L = \frac{1}{2}\dot{\lambda}\mathbf{M}\dot{\lambda}^{\top} + \frac{1}{2}\dot{\eta}^{\top}\mathbb{J}\dot{\eta} - \lambda\mathbf{MG} \tag{3}$$

$\mathbf{M}$ indicates a diagonal mass matrix, $\mathbf{J}$ is the inertia tensor, and $\mathbf{G} = [0\ 0\ g]^{\top}$ denotes gravity terms. Henceforth, consider $\lambda = [x\ y\ z]^{\top}$ and $\eta = [\phi\ \theta\ \psi]^{\top}$ as the position and attitude vectors, both expressed in the earth-fixed reference frame.

For controller design, the non-linear quadrotor dynamics can be written as

$$\begin{aligned}
m\ddot{x} &= -u\sin\theta + \xi_x \\
m\ddot{y} &= u\cos\theta\sin\phi + \xi_y \\
m\ddot{z} &= u\cos\theta\cos\phi - mg + \xi_z
\end{aligned} \tag{4}$$

unknown time-varying disturbances are represented by ξ_x, ξ_y, and ξ_z. On the other hand, disturbed rotational dynamics are given by

$$\mathbf{J}\ddot{\eta} = \boldsymbol{\tau}_\eta - \mathbf{C}(\dot{\eta}, \eta)\dot{\eta} + \boldsymbol{\xi}_\eta \tag{5}$$

with

$$\mathbf{J} = \begin{bmatrix} -I_x s_\theta & 0 & I_x \\ (I_y - I_z)c_\theta c_\phi s_\phi & I_y c_\phi^2 + I_z s_\phi^2 & 0 \\ I_z c_\theta^2 c_\phi^2 + I_y c_\theta^2 s_\phi^2 + I_x s_\theta^2 & (I_y - I_z)c_\theta c_\phi s_\phi & -I_x s_\theta \end{bmatrix}$$

$$\mathbf{C}(\dot{\eta}, \eta) = \begin{bmatrix} c_{11} & c_{12} & c_{13} \\ c_{21} & c_{22} & c_{23} \\ c_{31} & c_{32} & c_{33} \end{bmatrix} \tag{6}$$

and

$$c_{11} = (I_z - I_y)\dot{\psi}s_\phi c_\phi c_\theta^2$$

$$c_{12} = -I_x\dot{\psi}c_\theta + I_y(\dot{\theta}s_\phi c_\phi + \dot{\psi}c_\theta s_\phi^2 - \dot{\psi}c_\theta c_\phi^2) - I_z(\dot{\psi}c_\theta s_\phi^2 - \dot{\psi}c_\theta c_\phi^2 + \dot{\theta}s_\phi c_\phi)$$

$$c_{13} = 0$$

$$c_{21} = -I_x\dot{\psi}s_\theta c_\theta + I_y\dot{\psi}s_\theta c_\theta s_\phi^2 + I_z\dot{\psi}s_\theta c_\theta c_\phi^2$$

$$c_{22} = (I_z - I_y)\dot{\phi}s_\phi c_\phi$$

$$c_{23} = I_x\dot{\psi}c_\theta + I_y(-\dot{\theta}s_\phi c_\phi + \dot{\psi}c_\theta c_\phi^2 - \dot{\psi}c_\theta s_\phi^2) + I_z(\dot{\psi}c_\theta s_\phi^2 - \dot{\psi}c_\theta c_\phi^2 + \dot{\theta}s_\phi c_\phi)$$

$$c_{31} = \dot{\theta}I_x s_\theta c_\theta + I_y(-\dot{\theta}s_\theta c_\theta s_\phi^2 + \dot{\phi}s_\phi c_\phi c_\theta^2) - I_z(\dot{\theta}s_\theta c_\theta c_\phi^2 + \dot{\phi}s_\phi c_\phi c_\theta^2)$$

$$c_{32} = I_x\dot{\psi}s_\theta c_\theta - I_y(\dot{\theta}s_\theta s_\phi c_\phi + \dot{\phi}c_\theta s_\phi^2 - \dot{\phi}c_\theta c_\phi^2 + \dot{\psi}s_\theta c_\theta s_\phi^2) + I_z(\dot{\phi}c_\theta s_\phi^2 - \dot{\phi}c_\theta c_\phi^2 - \dot{\psi}s_\theta c_\theta c_\phi^2 + \dot{\theta}s_\theta s_\phi c_\phi)$$

$$c_{33} = -I_x\dot{\theta}c_\theta + (I_y - I_z)(\dot{\psi}c_\theta^2 s_\phi c_\phi)$$

For purposes of simplicity of the model representation, the shorthand notation for trigonometric functions is adopted [25], where $s_b = \sin b$ and $c_a = \cos a$. On the other hand, the control and disturbance torque vectors are denoted by $\boldsymbol{\tau}_\eta = [\tau_\phi\ \tau_\theta\ \tau_\psi]^T$ and $\boldsymbol{\xi}_\eta = [\zeta_\phi\ \zeta_\theta\ \zeta_\psi]^T$, respectively.

Since the quadrotor is an under-actuated non-linear system, two synthetic controllers are designed for tracking tasks of some desired reference position trajectory on the plane. For control design purposes, it is considered the output feedback errors given as follows

$$e_\mu = \mu - \mu^\star \tag{7}$$

for $\mu = x, y, z, \phi, \theta, \psi$. The superscript $\star$ stands for the desired reference trajectory. Moreover, in order to perform a proper motion control in X and Y directions, the desired pitch $\theta^\star$ and roll $\phi^\star$ references are computed from Equation (4) as follows

$$\theta^\star = \sin^{-1}\left(-\frac{1}{u}mv_x\right)$$

$$\phi^\star = \sin^{-1}\left(\frac{1}{u\cos\theta}mv_y\right) \tag{8}$$

To solve adequately the under-actuation problem, the angular dynamics needs to be faster than translational dynamics. In this way, the proposed motion controllers should be capable to lead the quadrotor to stable scenarios while performing a proper tracking of the planned references.

3. Syntheses of an Adaptive Robust Motion Controller

The syntheses of a novel adaptive robust motion controller is introduced in this section by using the robust control scheme introduced by the authors in [24]. In this proposal, it is improved the performance of the control scheme by reducing the high-gain effects and easing the tuning of the control parameters computed online by using the Bs-ANN.

3.1. Dynamic Compensators for Robust Control Design

In order to realize the stable control design, the quadrotor disturbed tracking error dynamics from Equations (4) and (5) are simplified as follows

$$\ddot{e}_\mu = v_\mu + \xi_\mu(t) \tag{9}$$

Moreover, $\xi_\mu(t)$ are assumed to be bounded time-varying disturbance signals locally approximated into a self-adaptive small interval of time around a given time instant $t_0 > 0$, say $[t_0, t_0 + \varepsilon]$, by r-th order Taylor polynomials as

$$\xi_\mu(t) \approx \sum_{n=0}^{r} \frac{\xi_\mu^{(n)}(t_0)}{n!}(t - t_0)^n = \sum_{n=0}^{r} \sigma_{n,\mu}(t - t_0)^n \tag{10}$$

where the superscript (n) stands for n-th order time derivative. Furthermore, to avoid velocity measurements, from Equation (9) structural estimates—known as integral reconstructors as well [26]—for time derivatives of velocity tracking errors are computed by

$$\widehat{e}_\mu = \int_{t_0}^{t} v_\mu \, dt \tag{11}$$

Here, initial conditions of the non-linear dynamic system, as well as the polynomial disturbance signal parameters are assumed to be completely unknown. Then, the polynomial relationship between integral reconstructors $\widehat{e}_\mu$ and actual velocity tracking error signals $\dot{e}_\mu$ is given by

$$\dot{e}_\mu = \widehat{e}_\mu + \sum_{n=0}^{r+1} \alpha_{n,\mu}(t - t_0)^n \tag{12}$$

where parameters $\alpha_{n,\mu}$ are assumed to be unknown as well.

In this fashion, the following family of controllers based on dynamic compensators to actively compensate polynomial disturbances can be synthesized as follows

$$v_\mu = -\beta_{r+3,\mu}\widehat{e}_\mu - \beta_{r+2,\mu}e_\mu - \delta_{r+1,\mu} \tag{13}$$

with

$$\dot{\delta}_0 = \beta_{0,\mu}e_\mu$$
$$\dot{\delta}_1 = \delta_{0,\mu} + \beta_{1,\mu}e_\mu$$
$$\vdots$$
$$\dot{\delta}_r = \delta_{r-1,\mu} + \beta_{r,\mu}e_\mu$$
$$\dot{\delta}_{r+1,\mu} = \delta_{r,\mu} + \beta_{r+1,\mu}e_\mu \tag{14}$$

Substitution of Equation (13) into Equation (9), closed-loop tracking error dynamics is then described by

$$e_\mu^{(r+4)} + \sum_{n=0}^{r+3} \beta_{n,\mu}e_\mu^{(n)} = 0 \tag{15}$$

Thus, closed-loop system stability criteria is fulfilled by selecting the control gains $\beta_{k,\mu}$ for $k = 0, 1, \ldots, r + 3$, such a way the characteristic polynomial of Equation (15) is stable (Hurwitz). By using the family of Hurwitz polynomials

$$P_{CL_\mu}(s) = (s + \gamma_\mu)^{r+4}, \quad \gamma_\mu > 0 \tag{16}$$

the control design parameters can be then computed by

$$\beta_{k,\mu} = \frac{(r+4)!}{k!(r+4-k)!}\gamma_\mu^{r+4-k} \tag{17}$$

In the present study, three layers B-spline artificial neural networks and particle swarm optimization are properly implemented to compute adaptive control gains in order to avoid possible undesirable high-gain control effects. Furthermore, first order Taylor polynomial expansions for approximation of disturbance signals are selected. Nevertheless, higher order polynomial expansions can be also chosen for applications where a much better approximation of disturbances is demanded. In this work, it is shown that first order polynomial disturbance approximations yield an acceptable motion trajectory tracking performance under significantly perturbed operating conditions.

Thus, from Equation (10), Taylor polynomial expansions for disturbance signals are described in this work as

$$\xi_\mu(t) \approx \sigma_{1,\mu} + \sigma_{2,\mu}(t - t_0) \tag{18}$$

where coefficients $\sigma_{1,\mu}$ and $\sigma_{2,\mu}$ are assumed to be uncertain. Moreover, the structural estimated variables and actual velocity tracking error signals are related by

$$\dot{e}_\mu = \widehat{\dot{e}}_\mu + \alpha_{0,\mu}(t - t_0) + \alpha_{1,\mu}(t - t_0)^2 \tag{19}$$

where parameters $\alpha_{i,\mu}$ are unknown.

In this sense, we proposed the following family of auxiliary controllers for robust quadrotor motion control

$$v_\mu = -\beta_{4,\mu}\widehat{\dot{e}}_\mu - \beta_{3,\mu}e_\mu - \beta_{2,\mu}\delta_{1,\mu} - \beta_{1,\mu}\delta_{2,\mu} - \beta_{0,\mu}\delta_{3,\mu} \tag{20}$$

with

$$\dot{\delta}_{1,\mu} = e_\mu$$
$$\dot{\delta}_{2\mu,} = \delta_{1,\mu}$$
$$\dot{\delta}_{3,\mu} = \delta_{2,\mu} \tag{21}$$

Thence, from Equations (9) and (20) the closed-loop error dynamics is governed by

$$e_\mu^{(5)} + \beta_{4,\mu}e_\mu^{(4)} + \beta_{3,\mu}e_\mu^{(3)} + \beta_{2,\mu}\ddot{e} + \beta_{1,\mu}\dot{e} + \beta_{0,\mu}e_\mu = 0 \tag{22}$$

The control gains $\beta_{k,\mu}$ for $k = 0, 1, \ldots, 4$ should be properly selected in order to the associated characteristic polynomials

$$P_{CL_\mu}(s) = s^5 + \beta_{4,\mu}s^4 + \beta_{3,\mu}s^3 + \beta_{2,\mu}s^2 + \beta_{1,\mu}s + \beta_{0,\mu} \tag{23}$$

are Hurwitz polynomials. In this fashion, reference trajectory tracking can be achieved:

$$\lim_{t \to \infty} e_\mu = 0 \quad \Rightarrow \quad \lim_{t \to \infty} \mu = \mu^\star \tag{24}$$

with μ and $\mu^\star$ standing for the real and planned references for translational and rotational trajectories, respectively.

Notice from (5) that the rotational dynamic model can be also be expressed as follows:

$$\ddot{\eta} = \mathbf{J}^{-1}\left(\tau_\eta - \mathbf{C}(\dot{\eta}, \eta)\dot{\eta}\right) + \mathbf{J}^{-1}\xi_\eta \tag{25}$$

which can be expressed matching the structure in (9). Therefore, from (21) it is observed that the synthetic controllers drive the system closed-loop dynamics. Finally, by analyzing the full non-linear dynamics, the control inputs nature and the robustness of the synthesized robust scheme, a suitable selection of the control inputs is given as follows

$$\begin{aligned}
u &= \frac{1}{\cos\phi\cos\theta}(mv_z + mg) \\
\tau_\psi &= I_z v_\psi \\
\tau_\theta &= I_y v_\theta \\
\tau_\phi &= I_x v_\phi
\end{aligned} \tag{26}$$

3.2. Adaptive Outline for Control Purposes

Adaptive control is a viable solution to avoid the use high-gain feedback or high frequency switching control actions for providing stability to many dynamic systems subjected to parametric uncertainty, unmodeled dynamics, and external disturbances [27]. In this work, we use a class of artificial neural networks for performing the tuning process of the control gains. The B-spline artificial neural networks are suitably integrated into adaptive motion controllers, where the tracking errors and their derivatives are used as the inputs of each network. The Bs-ANN functioning is based on the constant learning process of the physically system variables, therefore, have been successfully used to deal with system non-linear terms and uncertainty [28].

A B-spline function is a polynomial mapping defined by its extremes which uses a linear combination of the mono-variable and multi-variable basis functions. The B-spline networks, as depicted in Figure 2, are associative networks capable to adjust iteratively their synaptic weights for reproducing a specific function. The author in [29] proposes the following output:

$$y \;=\; \mathbf{aw}, \quad \mathbf{w} = \begin{bmatrix} w_1 \; w_2 \ldots w_h \end{bmatrix}^T, \quad \mathbf{a} = \begin{bmatrix} a_1 \; a_2 \ldots a_h \end{bmatrix} \tag{27}$$

where w_q y a_q are the q-th weight and the q-th basis function input, respectively; the quantity of synaptic weights is denoted by h. Each individual network output $y(t)$ is used in this work for computing dynamically the control gains. In this study, we introduce different experiments where the output of the neural networks differs: in experiments 1, 2, 4, and 5, it is computed just one control parameter while in experiment 3, three control parameters are computed by the adaptive scheme.

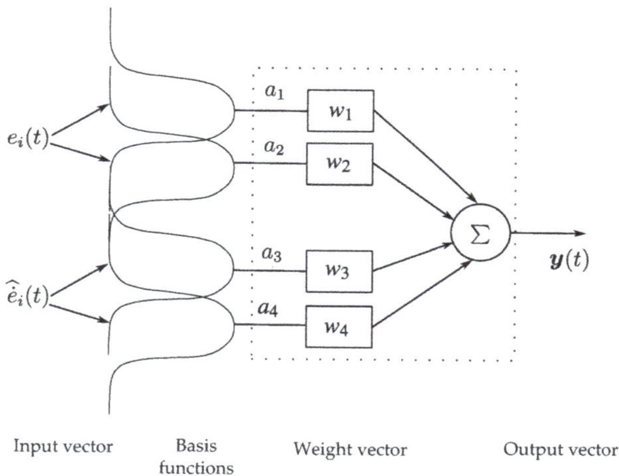

Figure 2. Three layer B-spline artificial neural network structure, Bs-ANN.

The actual output vector value minus the desired value defines the minimization error, which is used as the key term for the learning process. In this work, the following instantaneous learning rule has been adopted [30]

$$w_i(t) = w_i(t-1) + \frac{\ell e_i(t)}{\| \, \mathbf{a}(t) \, \|_2^2} a_i(t) \tag{28}$$

Here, ℓ represents the learning rate and $e_i(t)$ stands for the instantaneous output error. The adaptive process is achieved by the continuous training and the updating of the synaptic weights values considering the evolution of the inputs values. The Bs-ANN internal layer is constituted just by the basis functions, where the limits should be properly bounded by the adequate selection of the knot vector and basis function shape. In this proposal, four third order basis functions are employed for the adaptive scheme: two concerning the tracking error and two for the error derivative, as shown in Figure 2. It is important to mention that offline training of the Bs-ANN is performed for finding parameters during the adaptive scheme design process, in order to carry out the quadrotor to stable scenarios at the begging of the online training. To select properly these parameters, several quadrotor operational conditions are considered. The training data includes the transient and steady state response of the system, within the advantages of the proposed neural strategy, is that these data can come from an exact or approximate mathematical model or otherwise be measured data from the real system (inputs, outputs, control signals, etc.).

Thereafter, the control structure is summarized as follows: firstly, only the quadrotor position measurements are included as a feedback in order to determine the tracking errors. Later, integral reconstructors are suitably integrated for computation of the error derivatives which are used as inputs in the adaptive scheme for computing the control parameters and gains. Posteriorly, the virtual controllers v_x and v_y within the robust controller block are used for solving the under-actuation problem. Finally, the force and torque control inputs are injected to the system as variations of angular velocity of their four rotors, as portrayed in Figure 3.

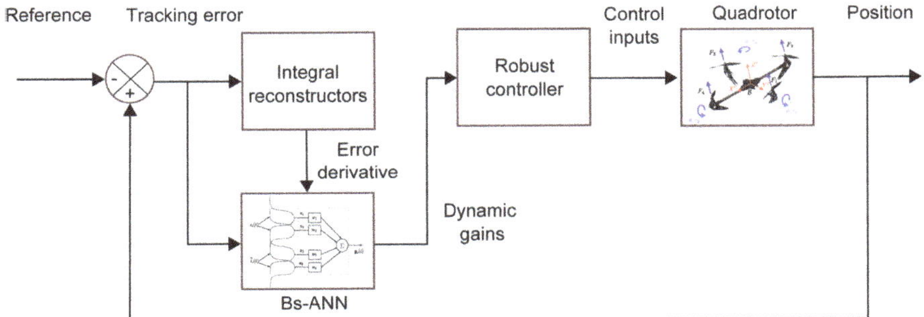

Figure 3. General structure of the adaptive robust motion control scheme.

4. Validation through Simulation Experiments

In this section, we investigate the applicability of the adaptive robust scheme for enhancing the tracking performance of a quadrotor non-linear system. Thus, several experiments are performed for an aerial vehicle numerically simulated. It is important to mention that the aim of the experiments is to portray some of the main contributions and advantages of implementing the proposed motion control strategy. Additionally, the experiments will demonstrate if the implementation of the proposal can be successfully extended for motion control of different types of autonomous vehicles. During the experiments, it is considered an aerial vehicle characterized by the set of parameters presented in Table 1.

Table 1. Parameters of the 6DOF non-linear quadrotor system.

Parameter	Units	Values
m	kg	0.98
g	m/s^2	9.81
l	m	0.25
J_x	kg m^2	0.012450
J_y	kg m^2	0.012450
J_z	kg m^2	0.024752

4.1. Polynomial Interpolation for Quadrotor Navigation

Bézier curves have been used widely and properly for path smoothing in robot navigation [31] and in motion control schemes for electric motors [32] and mechanical systems [33]. In the former, curves are expressed, such as parametric equations, where the time t is used to determine the values of coordinate pairs of (x, y) points graphed on the plane. In this work, a cubic Bézier curve is used and is defined by end points: (X_1, Y_1) and (X_4, Y_4), and control points: (X_2, Y_2) and (X_3, Y_3) such illustrated in Figure 4. In the second case, Bézier interpolation polynomials are suitably configured as position or velocity trajectory reference profiles, in order to soft the transition between two operation points for electromechanical and mechanical systems.

It is worthwhile to note that, due to its structure and after a proper selection of endpoints and control points, Bézier curves can be successfully implemented in a quadrotor to online computing the navigation path in cluttered environments, in order to ensure adequate obstacle avoidance manoeuvres while accomplishing a specific mission. On the other hand, it should be noted that derivatives of the trajectory references are not available in advance, and, in consequence, the proposed approach in this paper can be effectively implemented for this experiment.

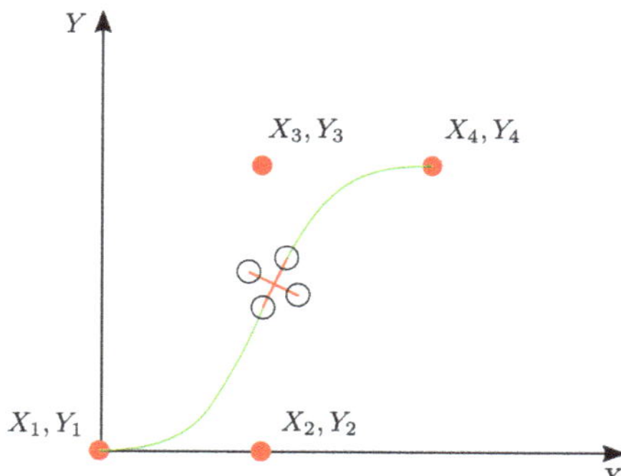

Figure 4. Cubic Bézier curve defined by a couple of pair of endpoints and control points.

During the first experiment, the quadrotor is tasked to perform the following: soft take-off to a height of 3 m; navigation through specific operation points in the space; and finally, soft landing, all of them by means of Bézier curves. It is worthwhile to note that the use of these curves is a viable strategy for solving properly the navigation and obstacle avoidance problems. Thus, in order to obtain smooth transitions between initial and final vertical operation points, the following motion scheme is adopted for take-off and landing tasks:

$$
z^\star = \begin{cases}
\Gamma_0 & 0 \leq t < T_1 \\
\Gamma_0 + (\Gamma_m - \Gamma_0)\mathcal{B}_z(t, T_1, T_2) & T_1 \leq t < T_2 \\
\Gamma_m & T_2 \leq t < T_3 \\
\Gamma_m + (\Gamma_0 - \Gamma_m)\mathcal{B}_z(t, T_3, T_4) & T_3 \leq t < T_4 \\
\Gamma_0 & t > T_4
\end{cases}
\tag{29}
$$

where $\Gamma_0 = 0$ and $\Gamma_f = 2$, given in meters, stand for the desired initial and maximum vertical positions. The time values given in seconds are as follows: $T_1 = 1$, $T_2 = 3$, $T_3 = 37$ and $T_4 = 40$. In addition, $\mathcal{B}_z$ is a Bézier polynomial [32] defined as

$$
\mathcal{B}_z(t, T_i, T_f) = \sum_{k=0}^{n} r_k \left(\frac{t - T_i}{T_f - T_i} \right)^k
\tag{30}
$$

with T_i and T_f as the initial and final transition times. Moreover, $n = 6$, and $r_1 = 252$, $r_2 = 1050$, $r_3 = 1800$, $r_4 = 1575$, $r_5 = 700$, $r_6 = 126$.

Subsequently, after the take-off, the rotorcraft is carry to desired positions in the horizontal plane, where the third order parametric equations used for navigation are defined as follows:

$$
\begin{aligned}
x^\star &= (1 - \mathcal{T})^3 X_1 + 3(1 - \mathcal{T})^2(\mathcal{T}X_2) + 3(1 - \mathcal{T})(\mathcal{T}^2 X_3) + \mathcal{T}^3 X_4 \\
y^\star &= (1 - \mathcal{T})^3 Y_1 + 3(1 - \mathcal{T})^2(\mathcal{T}Y_2) + 3(1 - \mathcal{T})(\mathcal{T}^2 Y_3) + \mathcal{T}^3 Y_4
\end{aligned}
\tag{31}
$$

Here, the values of the endpoints and control points are selected for performing a continuous navigation according to the parameters summarized in Table 2. Observe that four Bézier curves are used to define the whole navigation path and which is segmented for purposes of mathematical description.

Table 2. Control and endpoint values for the Bézier curves.

Segment	Time Lapse [s]	$\mathcal{T}$	X_1	Y_1	X_2	Y_2	X_3	Y_3	X_4	Y_4
1	$0 \leq t < 10$	$\frac{t}{10}$	0	0	1	0	1	2	2	2
2	$10 \leq t < 20$	$\frac{t}{10} - 1$	2	2	3	2	3	4	4	4
3	$20 \leq t < 30$	$\frac{t}{10} - 2$	4	4	5	4	5	2	6	2
4	$t \geq 30$	$\frac{t}{10} - 3$	6	2	7	2	7	0	8	0

On the other hand, external vibrating disturbance forces have been included after 12 s for robustness assessment purposes of the introduced motion control scheme, and are given by

$$\xi_j = \mathcal{A}_j \sin(\omega_j t) \tag{32}$$

with $j = x, y, z$, $\mathcal{A}_x = \mathcal{A}_y = 1\,\text{N}$, $\mathcal{A}_z = 2\,\text{N}$, and $\omega_x = \omega_y = \omega_z = 10\,\text{rad/s}$.

In Figure 5, it is presented the quadrotor flight performance by implementing the proposed controller, where a proper path following is exhibited. Throughout the manuscript, the use of solid and dashed lines for representing real and desired trajectories is adopted, respectively. As observed in Figure 6, the Bézier curves are successfully implemented for navigation between operation positions, and as a consequence of the proposed controller, a proper trajectory tracking of the planned references is achieved. Moreover, according to this figure, angular tracking of the online computed references $\phi^\star$ and $\theta^\star$ is achieved in spite of there is not information about the derivatives of these references since a properly integration of integral reconstructors and neural networks within the robust motion control approach is achieved.

Furthermore, it is evident the satisfactory performance of the quadrotor tracking motion control scheme even though the quadrotor is subjected to undesired harmonic forces. Notice that regulation around $\psi^\star = 0$ rad is performed in this experiment. Additionally, Figure 7 portrays the controlled vertical quadrotor dynamics, the height control, and yaw motion regulation. From this figure, the utility of the Bézier polynomial curve, where a soft take-off and landing are achieved thanks to the mathematical framework introduced by Equations (30) and (52) is appreciated. In the next section, the ground effect is included within the analysis in order to assess the control scheme robustness for controlling the quadrotor vertical motion.

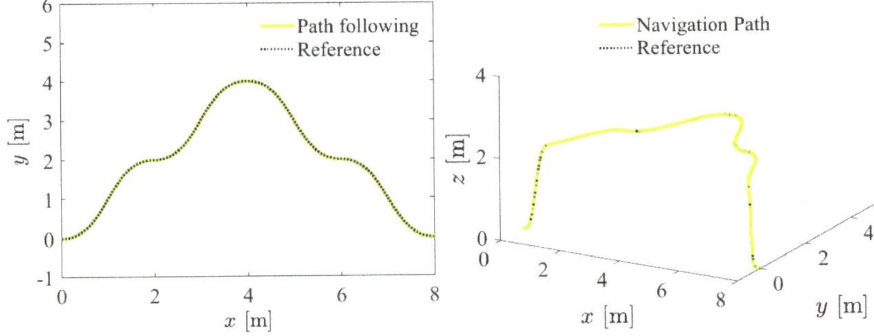

Figure 5. Quadrotor navigation on the plane and space.

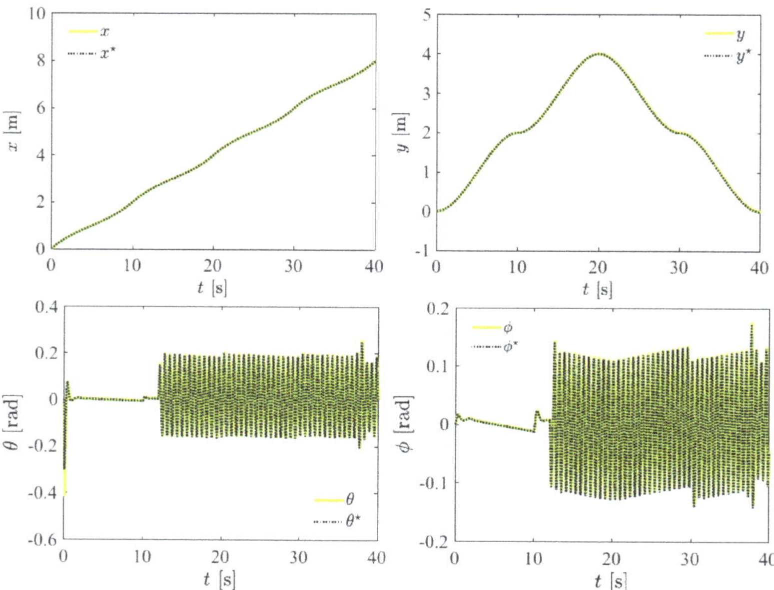

Figure 6. Lateral and longitudinal motion tracking in experiment 1.

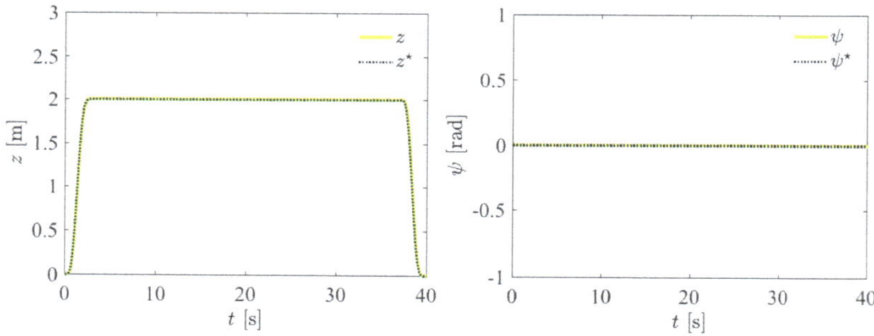

Figure 7. Vertical motion tracking for experiment 1.

For this experiment the following desired Hurwitz polynomial has been selected,

$$P_d(s) = (s + \gamma^2)^5 \tag{33}$$

where, in order to ensure close-loop stability and the properly tracking of the planned trajectory, the control gains in (23) should match the following

$$
\begin{aligned}
\beta_{4_i} &= 5\gamma_i \\
\beta_{3_i} &= 10\gamma_i^2 \\
\beta_{2_i} &= 10\gamma_i^3 \\
\beta_{1_i} &= 5\gamma_i^4 \\
\beta_{0_i} &= \gamma_i^5
\end{aligned}
\tag{34}
$$

where γ_i, for $i = x, y, z, \phi, \theta, \psi$, is the unique online computed control parameter. To improve and ease the parameter selection process in this experiment, each of these control parameters are suitably derived by the adaptive framework introduced in Figure 2, where the output of each individual neural network is the value for the control parameter γ_i. As it is presented in Figure 8, dynamical updating, as well as a successful parameter computation of the control gains, is achieved by using the adaptive B-spline artificial neural networks.

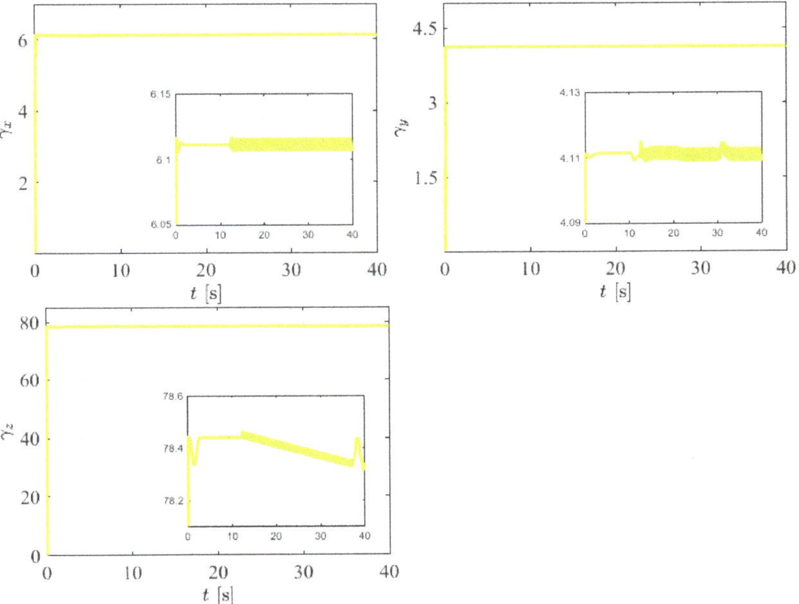

Figure 8. Adaptive γ_i control parameters, for $i = x, y, z$.

In Figure 9, it has been included results considering both perturbed and unperturbed cases in order to contrast the compensation action of the adaptive robust control scheme. It is worthwhile to note, from Figure 9b, that it is possible to track, satisfactorily, the references, as well as being demonstrated in Figure 9a. Nevertheless, the vibrating disturbance compensation is not present in the unperturbed case. By analyzing Figure 9b, it is evident the reachability of the control commands which benefits the non-saturation of the actuators. It is also important to mention that similarly as the oscillations due to the control compensation action, in Figure 6 it is appreciated the compensation of the vibrating disturbance forces affecting translational dynamics since are related with the rotational trajectory tracking trough the under-actuation property.

According to the results, the proposed control method is robust and able to efficiently reduce induced oscillations. Additionally, it is demonstrated that Bézier polynomial interpolation can be widely and satisfactorily exploited in quadrotor motion control systems: path and trajectory tracking. The experiment presented in this section illustrates that the complex quadrotor non-linear system is motion controlled in an acceptable way. As no information is required about derivatives of the trajectory references and from the external disturbances the control process is simplified significantly.

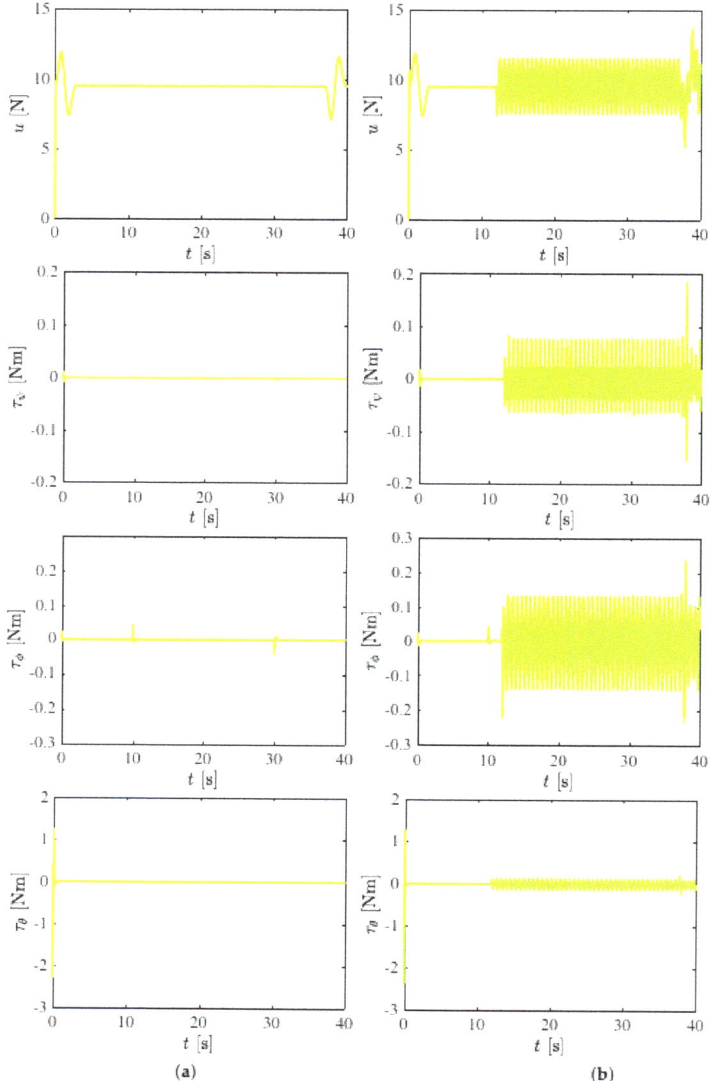

Figure 9. Computed control inputs in experiment 1. (**a**) Unperturbed. (**b**) Perturbed.

4.2. Improved Robust Quadrotor Autonomous Landing

One of the most essential requirements for a VTOL vehicle is to ensuring a safe landing flight phase. Rotorcraft are subjected to significant variations in motion control during take-off and landing stages due to the increase in lift force when they are close to the ground. Such phenomena are known as the ground effect [34]. The aim of this experiment is to assess the capabilities of the proposed controller for dealing with the ground effect in simulation. Therefore, the Cheeseman and Bennett modified ground effect model, proposed for quadrotors by authors in [35], are used, which state the following:

$$\frac{u}{u_r} = 1 - \rho\left(\frac{r}{4z_r}\right)^2 \tag{35}$$

where the ratio $\frac{u}{u_r}$ is equal to one outside of the ground-effect. In addition, r is the propeller radius, z_r represents the distance from the rotor to the ground, u and u_r is the input thrust commanded and the generated real thrust, respectively. Notice, the third expression of equations set (4) is affected by the introduced model representation of the ground effect phenomenon, where it is evident that

$$u_r = u + u_r \rho\left(\frac{r}{4z_r}\right)^2 \tag{36}$$

and referring to the above equation and using the real generated input thrust in the nominal mathematical model it yields the following

$$m\ddot{z} = u_r \cos\theta\cos\phi - mg \tag{37}$$

or

$$m\ddot{z} = u\cos\theta\cos\phi + u_r\rho\cos\theta\cos\phi\left(\frac{r}{4z_r}\right)^2 - mg \tag{38}$$

Thereafter, without loss of generality

$$m\ddot{z} = u\cos\theta\cos\phi - mg + \xi_z \tag{39}$$

with

$$\xi_z = u_r\rho\cos\theta\cos\phi\left(\frac{r}{4z_r}\right)^2 \tag{40}$$

where ξ_z should be compensated by the adaptive robust motion control approach. In addition, the following data have been used during the simulation: $\rho = 10$, $r = 0.1$ m, and $z_r = 0.1$ m.

On the other hand, in Figure 10 the quadrotor landing is illustrated. Here, it is used two different values for the learning rate ℓ and for the weighting vector for vertical motion $\mathbf{w}_z = [w_{1,z}, w_{2,z}, w_{3,z}, w_{4,z}]$, in order to illustrate two cases where the effect of increasing or decreasing the parameter values within the adaptive framework defines the quadrotor operation. Moreover, it is observed that a better tracking performance of the closed-loop system is achieved when a suitably selection of the parameters is done. In Table 3 are showcased the respective values for the aforementioned parameters in each case.

It is relevant to mention that in this experiment it is adopted the same set up outlined by expressions (33) and (34) defined in the previous section. Thus, as corroborated by the dynamic behavior of γ_z in Figure 10, online computation of the control parameters is accomplished dynamically by the adaptive framework. From the same figure, it is also appreciated that the magnitude of the control effort is modified in function of the disturbance force exerted as a consequence of the ground effect. Nevertheless, a significant deviation of the actual motion from the planned reference is observed in the first case. In contrast, in the second case, acceptable attenuation levels of induced oscillations is attained by a proper selection of the parameters presented in Table 3.

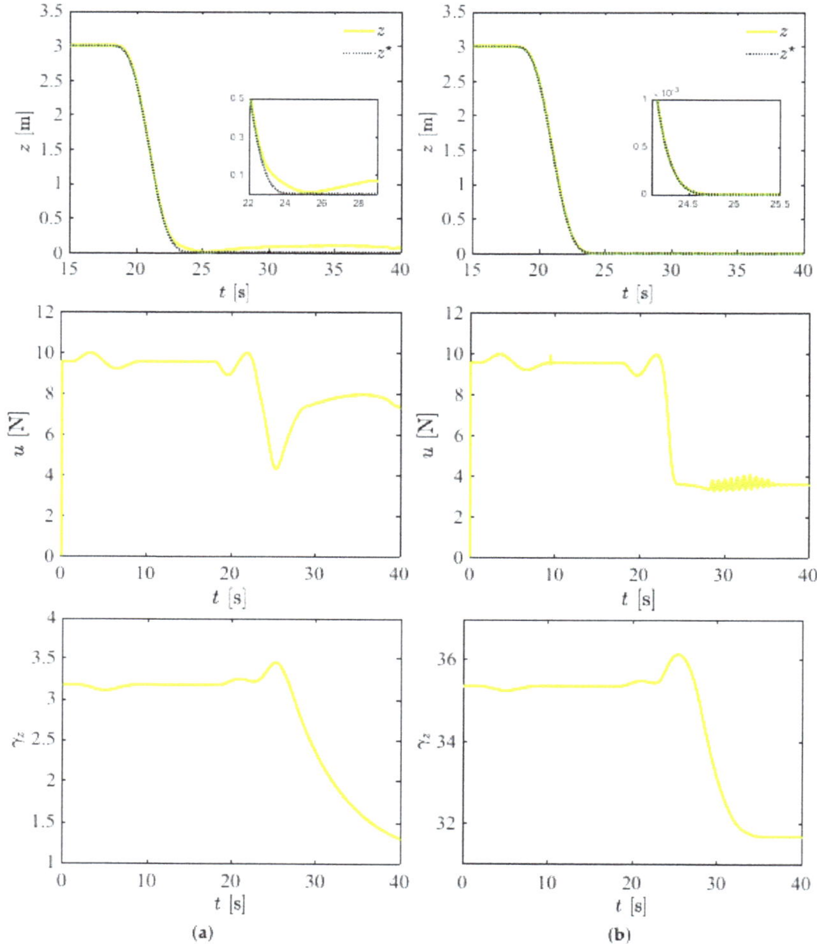

Figure 10. Quadrotor autonomous landing under the ground effect phenomenon: (**a**) First case. (**b**) Second case.

The key for a successfully performance of the adaptive scheme depends on a properly selection of the adaptive parameters during the design process. Note that the selection of the initial weights within the offline training procedure, different operational conditions can be take into account for improving the initial system response, and, in this way, leading the quadrotor non-linear system to stable scenarios. In the next section, a different setup is introduced for selection of the control parameters: a desired Hurwitz polynomial where three parameters will be computed and a optimized selection by means of particle swarm theory.

Table 3. Parameters for the adaptive framework in experiment 2.

Case	ℓ_z	$w_{1,z}$	$w_{2,z}$	$w_{3,z}$	$w_{4,z}$
First	5×10^{-9}	1	1	2	1
Second	5×10^{-4}	30	20	3	3

4.3. Bs-ANN Offline Training by Particle Swarm Optimization

Inspired by the social behavior observed in fish schools and bird flocks, particle swarm optimization (PSO) has been proposed as an effective solution for solving a wide range of optimization problems [36]. The use of intelligent agents, called particles, allows this algorithm to iteratively find the best solution on a defined space of searching. For this reason, potentials of PSO has been properly exploited in different engineering and researching applications, such as tuning of automatic controllers [37] and artificial neural networks training [38]. In the second experiment, the PSO is used for the offline training of the BsNN (selection of the initial weights). The training process is performed while the system is commanded to reach a step reference for vertical translational motion, where the closed-loop response information is used for designing the objective function f_o to be minimized.

Figure 11 portrays the closed-loop response of a second order system. Here, it can be observed that there exist several parameters can be used in the design of the objective function in order to minimize the tracking error and the control efforts: t_r, t_s, M_p, and t_p stand for the rise time, settling time, maximum peak, or overshoot and peak time, respectively.

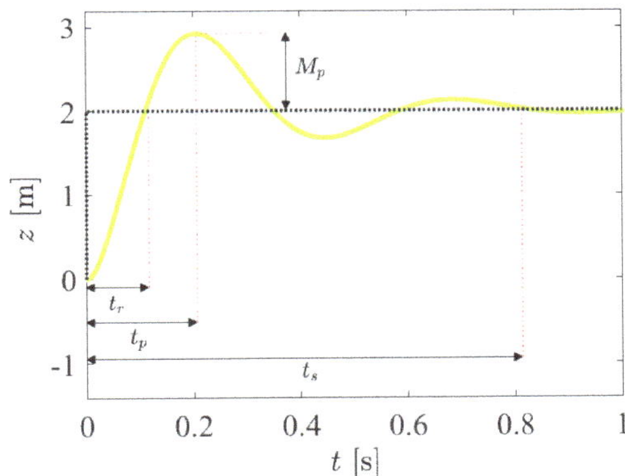

Figure 11. Time response of a closed-loop controlled second order dynamical system.

In this study, only the overshoot data are used as design parameter of the following objective function

$$f_o = \epsilon (M_p + \text{ITAE}) + \kappa (\text{ISCI}) \tag{41}$$

where the coefficients $\epsilon = 0.5$ and $\kappa = 0.1$ penalize the error and the magnitude of the control inputs, respectively. On the other hand, the integral time absolute error (ITAE) index is computed as follows

$$\text{ITAE} = \int_0^t t|e_z|\, dt \tag{42}$$

here e_z is the tracking vertical error and t is the time variable. Additionally, the integral squared control input (ISCI) term is introduced in Equation (43).

$$\text{ISCI} = \int_0^t u^2\, dt \tag{43}$$

In contrast with the previous experiments, it has been selected the following Hurwitz polynomial:

$$P_d(s) = (s^2 + 2\zeta_c \omega_c s + \omega_c^2)^2 (s + P_c) \tag{44}$$

here $\omega_{n_c}, \zeta_c, P_c > 0$, are the controller adjustment parameters. Therefore, concerning Equation (22), the control gains can be selected as follows for ensuring closed-loop stability and the properly tracking of the planned trajectory

$$\beta_{4_z} = 4\zeta_c\omega_c + P_c$$
$$\beta_{3_z} = 2\omega_c^2 + 4\zeta_c^2\omega_c^2 + 4P_c\zeta_c\omega_c$$
$$\beta_{2_z} = 4\omega_c^3\zeta_c + 2P\omega_c^2 + 4P_c\zeta_c^2\omega_c^2$$
$$\beta_{1_z} = 4P_c\omega^3\zeta_c + \omega_c^4$$
$$\beta_{0_z} = P_c\omega_c^4 \tag{45}$$

For the third experiment, the quadrotor take-off stage is analyzed. In order to improve and ease the tuning process, the control parameters are properly computed online by using artificial neural networks which trained offline by a PSO framework.

In the Algorithm 1, it is presented the pseudocode for the training process, where a simulation time of 10 seconds is adopted.

Algorithm 1: Evaluation of the objective function f_o.

    Input: $\mathbf{w}_z = 0$ `// 1 × 4 weight vector`
    Output: f_o

1 $\epsilon = 0.5$, $\kappa = 0.1$, $\Delta t = 0.001$ `// constants`
2 $z(0) = 0$, $\dot{z}(0) = 0$, $t_s = 0$ `// initial conditions`
3 **for** $k = 1$ **to** s **do**   `// s = `$\frac{t_s}{\Delta t}$`, `$t_s = 10$
4   Calculus of ζ_c ω_c and P_c
5   Solution of system dynamics   `// by Runge-Kutta Fehlberg Numerical`
        method
6      $t_s = t_s + \Delta t$
7   Storing $\mathbf{e}[k] = e_z$, $\mathbf{u}[k] = u$, $\mathbf{t}[k] = t_s$
8 **end**
9 $n = s$
10 Get M_p `// by means of stepinfo MATLAB function`
11 Calculus ITAE $= \sum_1^n \mathbf{t}|\mathbf{e}|^2 \Delta t$
12 Calculus ISCI $= \sum_1^n |\mathbf{u}|^2 \Delta t$
13 Evaluation of f_o

The MATLAB optimization toolbox is used for the execution of the PSO algorithm. It is worthwhile to note that the procedure in Algorithm 1 is evaluated in each iteration of the optimization process, in order to determine the best set of control parameters who minimizes the objective function, which has been designed in function to the vertical motion tracking error, as well as the control input effort. Moreover, for this simulation experiment the PSO algorithm is configured with the dimensions of the search space defined by the low and upper boundaries $lb = -5$ and $ub = 5$, respectively, and a swarm size of 50 particles.

Additionally, in order to highlight the performance of the introduced novel adaptive robust control strategy, in this section it is illustrated the applicability of offline training of Bs-ANN neural networks by description of two relevant scenarios: in the first the offline training is carried out for determining initial values of control parameters ζ_c ω_c and P_c without using the online learning. On the other hand, online training is considered for computation of the parameters values throughout second scenario. Henceforth, we identified the scenarios, respectively, as fixed and adaptive. The yielded results are portrayed in Figures 12 and 13.

It is worth to mention that from Figure 12 it is observed that the performance for both scenarios looks similar. Nevertheless, the control signal efforts and the error are significantly decreased by using the adaptive strategy. The ISCI and the ITAE indexes are used also for a quantitative comparison and is summarized in Table 4 for both cases in experiment 3.

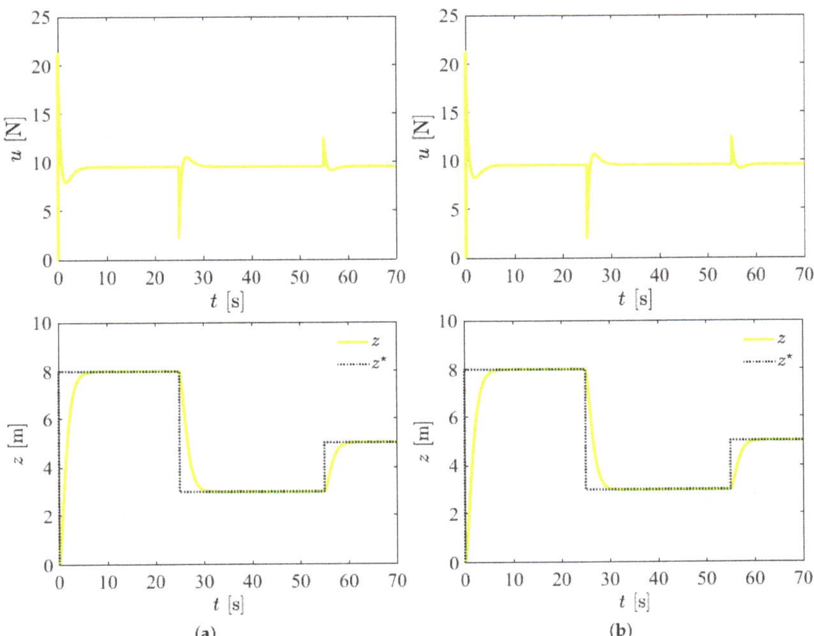

Figure 12. Tracking motion for experiment 3: (**a**) First scenario. (**b**) Second scenario.

Table 4. Computed ISCI in experiment 3.

Gain Case	ISCI	ITAE
Fixed	6.4181×10^3	423.9004
Adaptive	6.4140×10^3	422.9049

It is important to point out that in first scenario it is also achieved an acceptable performance of the introduced control approach. The tuning procedure of the control gains in automatic control systems is not always an easy tasks since it depends on the designer experience for selecting the control gains. Thus, after a properly setup of the PSO scheme, it is possible to ease the tuning process where several control gains or parameters need to be selected: five gains in the present study. Moreover, in Figure 12, we highlighted the useful of the offline and online training process in the quadrotor motion control. Here, large overshoot and oscillation is avoided from the the closed-loop response by an efficient implementation of the adaptive robust motion control strategy.

On the other hand, in Figure 13, it can be appreciated the effects for using the online training of the Bs-ANN in contrast with the fixed case utilized in the first scenario of third experiment. According to the information presented in this figure, it is corroborated that by using the full adaptive scheme it is possible to improve the closed-loop response of the quadrotor system by suitably adjusting the control parameters. Notice that the introduced control scheme, it is able to perform efficiently regulation and trajectory tracking tasks even though there is not full knowledge of the non-linear quadrotor mathematical model, as well as the external vibrating disturbances.

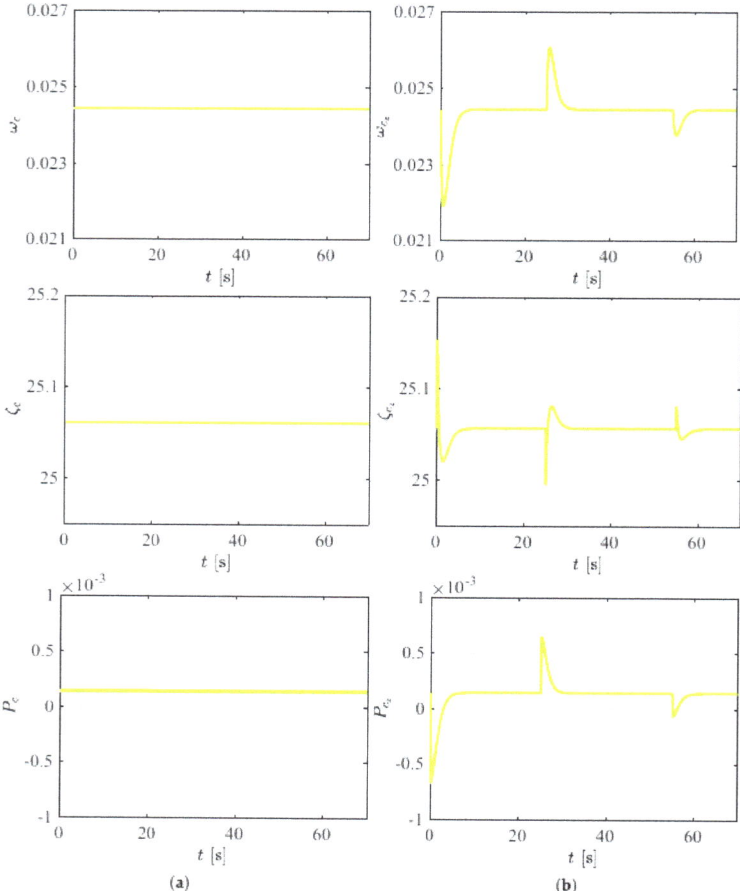

Figure 13. Fixed and adaptive control parameters used in experiment 3: (**a**) First scenario. (**b**) Second scenario.

4.4. Quadrotor Subjected to Wind Gust Disturbances

In this section, a Dryden wind gust model is used for the assessment of control robustness. From the set of Equation (4), it is evident that in presence of induced disturbance torques, the angular, as well as the translational trajectory tracking, will be deteriorated. Therefore, in this experiment, the quadrotor is disturbed while it is hovering and path following in order to simulate different scenarios which it would usually face within a wide range of applications. Consider the wind gust mathematical model [39] given by

$$\zeta_\eta(t) = d_w^s + \sum_{\sigma=1}^{n} \mathcal{A}_\sigma \sin(\omega_\sigma t + \varphi_\sigma) \tag{46}$$

for $\eta = \phi, \theta, \psi$. Expression (46) considers that the disturbance caused by wind field is proportional to the wind speed [39], which is described as a family of time-varying excitations. On the other hand, ω_σ and φ_σ are randomly selected frequencies and phase shifts, respectively; n is the number of sinusoids, $\mathcal{A}_\sigma$ is the amplitude, and d_w^s is a static term for wind disturbance. Thus, the mathematical expression in (46) can be integrated in

(5) for this simulation experiment as torque disturbances ζ_ϕ, ζ_θ and ζ_ψ, with $n = 6$ for ϕ and θ, and $n = 7$ for ψ. Disturbance parameters are summarized in Table 5.

Table 5. Values for simulated torque disturbances.

ξ_η	d_w^s	$\mathcal{A}_1 \dots \mathcal{A}_n$	$\omega_1 \dots \omega_n$	$\varphi_1 \dots \varphi_n$
ζ_ϕ	0.3	0.27, 0.45, 0.06, 0.45, 0.3, 0.15	$\pi(2.5, 2, 0.4, 0.08, 0.07, 0.05)$	$-1.2, 2.7, -9.5, 1, 0.5, 2$
ζ_θ	0.6	0.2, 0.1, 0.4, 0.1, 0.2, 0.1	$\pi(1.5, 2, 0.4, 0.03, 0.07, 0.05)$	$-0.3, 1.7, -1.5, 1, 1.5, 0.3$
ζ_ψ	2	0.5, 0.725, 1, 0.5, 0.25, 0.5, 0.25	$\pi(2.5, 2, 0.4, 0.2, 0.008, 0.07, 0.05)$	$-3, 7, -9.5, 0, 1, 1.5, 2$

Consider the following planned references for lateral and longitudinal quadrotor motion in this experiment,

$$x^\star = 5\cos(T) + \cos(3T)\cos(T)$$
$$y^\star = 5\sin(T) + \cos(3T)\sin(T) \tag{47}$$

with $T = 0.1t$, and the Bézier based motion profile for vertical motion defined by Equations (30) and (52) with the following data: $\Gamma_0 = 0$, $\Gamma_f = 5$, $T_1 = 2$, $T_2 = 10$, $T_3 = 57$ and $T_4 = 65$. Additionally, the yaw motion is regulated about a constant angle $\psi^\star = 0$ rad. Soft transition between initial condition and the regulation point is accomplished by a Bézier polynomial.

Figures 14 and 15 describes the effective performance of the adaptive robust motion control scheme (20), which compensates the disturbance forces induced by the wind gust model introduced in (46). Moreover, it is evident excellent levels of oscillations attenuation by using our control approach.

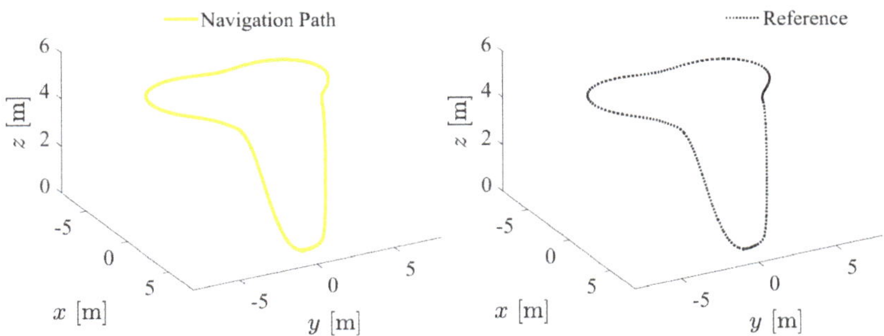

Figure 14. Trajectory tracking for the experiment 4.

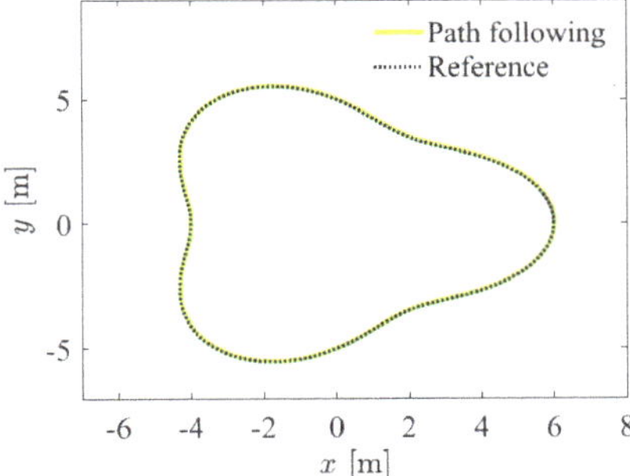

Figure 15. Path following on the plane for the experiment 4.

From Figure 16, it is observed that the quadrotor is able to efficiently perform trajectory tracking tasks in spite of there is not previous information of the disturbance torques while tracking the planned references introduced in (47). On the other hand, the control input forces and torques generated by the proposed controller are presented in Figure 17. Here, it is appreciated a properly compensation of the disturbance effects which by the computed control inputs. The closed-loop system response for rotational dynamics is plotted in Figure 18, where it is corroborated an efficient performance of the introduced adaptive robust control approach, as done in previous experiments. It is important to mention that during experiment 4 it is adopted the same process for the computation of the control gains in the first experiment.

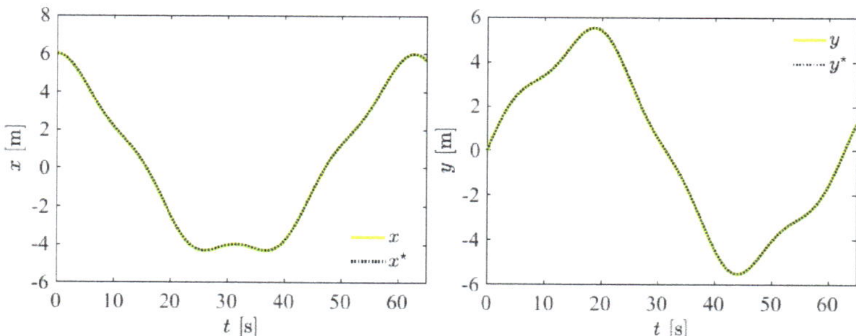

Figure 16. Translational motion tracking for experiment 4.

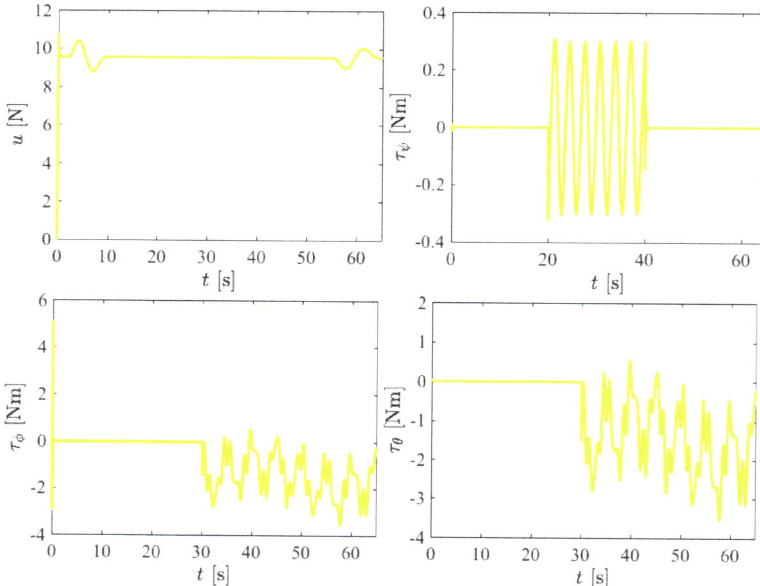

Figure 17. Computed control inputs in experiment 4.

Finally, in Figure 18, it can be seen the control parameters for rotational motions, which are computed online by means of the adaptive BS-ANN scheme. In addition, it is corroborated that even though there is not available information from derivatives of the angular references, the under-actuation problem is properly solved by the use of the neural networks and the integral reconstructors, thereby a good tracking of the online computed references $\phi^\star$ and $\theta^\star$ is achieved.

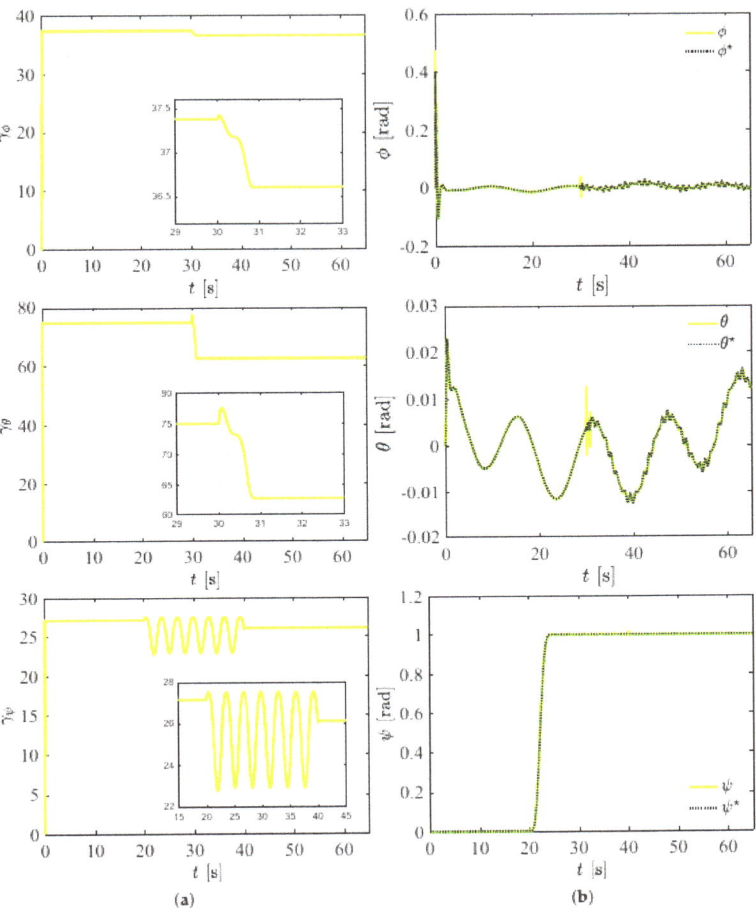

Figure 18. Dynamic rotational closed-loop response for the experiment 4: (**a**) Computed rotational adaptive control parameters. (**b**) Angular tracking motion.

4.5. Robustness against Uncertainty of Quadrotor Mass

Another important issue for controlling a quadrotor is the variations of the nominal mass. Notice that the online computed references in (8) which define a proper motion on the plane depends on the nominal mass. Therefore, the quadrotor is supposed to follow the references considering the nominal mass value. Thus, during this experiment, it is probed if the vehicle flight may be deteriorate significantly when an extra mass is added.

Consider the following mass variation for this experiment

$$m = m_n + \mathcal{M}_\Delta(t) \tag{48}$$

where $m_n = 0.973$ stands for the nominal quadrotor mass in kg, and $\mathcal{M}_\Delta(t)$ is an abrupt change of the mass quadrotor described by a modified impulse function given by

$$\mathcal{M}_\Delta(t) = \begin{cases} \left(\dfrac{1}{\sqrt{2\pi\sigma_\Delta^2}}\right) e^{-\left(\frac{(t-a)^2}{2\sigma_\Delta^2}\right)} + b & 0 \leq t < 20.4 \\[3mm] \left(\dfrac{1}{\sqrt{2\pi\sigma_\Delta^2}}\right) e^{-\left(\frac{(t-c)^2}{2\sigma_\Delta^2}\right)} + d & t \geq 20.4 \end{cases} \tag{49}$$

with $a = 20.25$, $b = 0$, $c = 19.55$, $d = 0.7$, and $\sigma_\Delta = 0.4$. During the experiment it is adopted a spiral shape planned reference given by the next parametric equations

$$\begin{aligned} x^\star &= r\cos(T) \\ y^\star &= r\sin(T) \end{aligned} \tag{50}$$

where

$$r = 5 + 0.2\cos(t) \tag{51}$$

and $T = \frac{t}{6}$. Inspecting Figure 19, it is appreciated that the abrupt variation in the quadrotor mass does not affect significantly the following of the planned reference. In Figure 20, it is observed a slightly deviation of the quadrotor angular tracking in contrast with the nominal references $\theta_n^\star$ and $\phi_n^\star$, computed with the nominal mass. Moreover, it is evident that after a brief period of time the quadrotor is able to recover from the perturbation and perform a proper tracking of the desired references thanks to the robustness of the proposed control scheme.

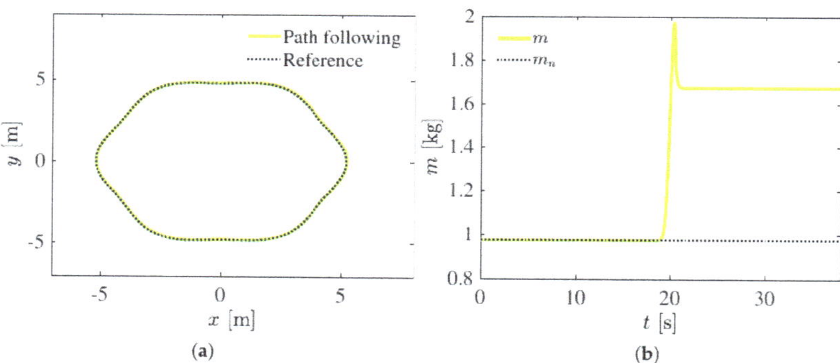

Figure 19. Simulation results for experiment 5. (**a**) Path following. (**b**) Mass variation given by Equation (48).

In Figure 21a, it is presented the compensation to the mass variation at 20.4 s by the control input u. Figure 21b portrays the vertical motion which is performed before the path following, and is given by the following Bézier polynomial

$$z^\star = \begin{cases} \Gamma_0 & 0 \leq t < T_1 \\ \Gamma_0 + \left(\Gamma_f - \Gamma_0\right)\mathcal{B}_z(t, T_1, T_2) & T_1 \leq t < T_2 \\ \Gamma_f & t \geq T_2 \end{cases} \tag{52}$$

where $\Gamma_0 = 0$ and $\Gamma_f = 5$, given in meters, stand for the desired initial and maximum vertical positions. The time values given in seconds are as follows: $T_1 = 2$, $T_2 = 10$. In addition, $\mathcal{B}_z$ is the Bézier polynomial introduced in (30) with T_i and T_f as the initial and final transition times. Moreover, $n = 6$, and $r_1 = 252$, $r_2 = 1050$, $r_3 = 1800$, $r_4 = 1575$, $r_5 = 700$, $r_6 = 126$.

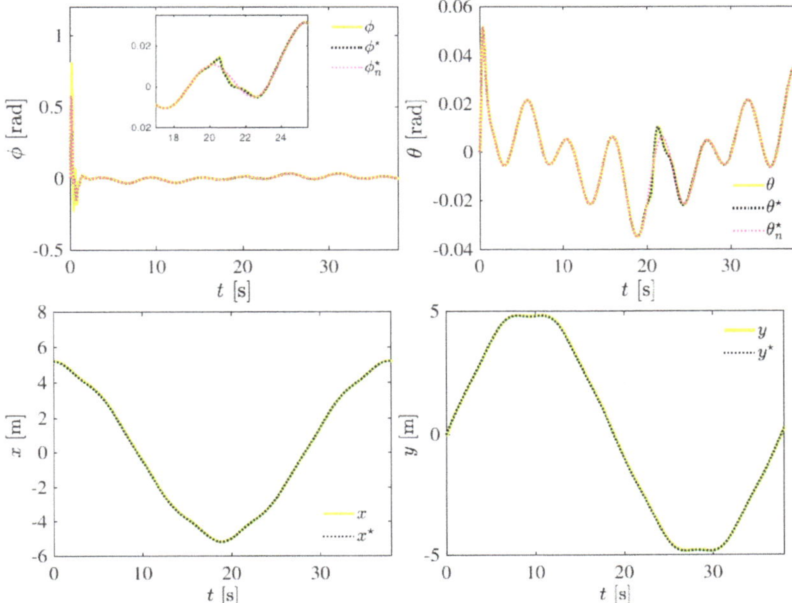

Figure 20. Reference tracking results for experiment 5.

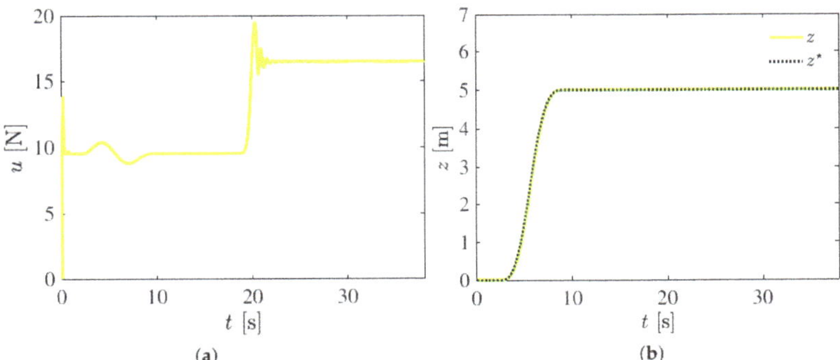

Figure 21. Simulation results for experiment 5. (**a**) Main force control input. (**b**) Controlled vertical motion.

Future studies will address other methodologies for trajectory generation. Interested readers are referred to [40,41] and references therein for further information on trajectory generation. Moreover, algebraic estimators [42] will be explored for determining variation in the quadrotor nominal mass due to unknown payload and damage or failure in the quadrotor frame.

4.6. Discussion of Results

Throughout the presented experiments in this work, it is corroborated that the introduced control scheme is able to properly leads the quadrotor to stable scenarios in spite of the presence of external forces and torque disturbances, as well as mass uncertainty. Different from the important adaptive robust contributions proposed in [21–23], in our approach disturbance observers are unnecessary, only positions measurements are required for motion control, and the use of high frequency discontinuous control actions, as a consequence of the use of signum functions, is avoided. Additionally, due to the low dependency of the quadrotor non-linear mathematical model, the robustness against external disturbances and since it is not required the error derivatives for the implementation of our control scheme, we widely recommend this as an viable strategy for controlling other autonomous systems.

5. Conclusions

A novel adaptive robust neural motion control scheme for quadrotor systems has been introduced in this study. The proposed controller guarantees that the desired trajectory can be tracking by the quadrotor vehicle. The control framework is composited by using dynamic compensators along with an adaptive strategy based on B-spline artificial neural networks. Moreover, an artificial intelligence mechanism based on PSO theory was properly employed for improving the design of the control strategy, as well as the dynamic closed-loop response of the quadrotor system, where ITAE and ISCI metric indexes have been used for measuring the control performance. The introduced research is particularly important because of its potential application in motion tracking control of quadrotor systems, where the use of neither tracking error derivatives nor disturbance observers is required. Several simulation experiments were proposed for purposes of performance and robustness assessment. It was corroborated that, by using the introduced control scheme, the quadrotor dynamic response is sufficiently robust for driving the system to stable scenarios. Vibrating disturbance forces and torques, uncertainty of the quadrotor mass, and wind gusts affecting the quadrotor stable motion were used for testing the control scheme robustness. The suitable integration of adaptive and robust control allows to compensate external disturbances during the quadrotor navigation and soft-motion during take-off and landing. Additionally, the quadrotor performance is improved significantly in contrast when it is adopted fixed values for the control gains. The obtained results, thus, prove that the controlled quadrotor system is able to achieve acceptable control accuracy levels for both trajectory tracking and path following in spite of been subjected to undesired disturbances. Finally, it is worth pointing out that the proposed control strategy can be further extended for the motion control of different autonomous systems subjected to external vibrating disturbances, which will be addressed in future works.

Author Contributions: Research, H.Y.-B., F.B.-C., R.T.-O., A.F.-C., C.S., D.S.; Methodology, H.Y.-B., F.B.-C., R.T.-O., C.S., D.S.; Software, H.Y.-B.; Formal analysis, H.Y.-B., F.B.-C., R.T.-O., A.F.-C.; Supervision, H.Y.-B., F.B.-C., R.T.-O., A.F.-C.; Writing, H.Y.-B., F.B.-C., R.T.-O., A.F.-C., C.S., D.S., Project administration, H.Y.-B. All authors have read and agreed to the published version of the manuscript.

Funding: This research received no external funding.

Institutional Review Board Statement: Not applicable.

Informed Consent Statement: Not applicable.

Data Availability Statement: Not applicable.

Conflicts of Interest: The authors declare no conflict of interest.

References

1. Castillo, P.; Dzul, A. Aerodynamic Configurations and Dynamic Models. In *Unmanned Aerial Vehicles*; Lozano, R., Ed.; John Wiley & Sons, Ltd.: Hoboken, NJ, USA, 2013; pp. 1–20.
2. Leutenegger, S.; Hürzeler, C.; Stowers, A.K.; Alexis, K.; Achtelik, M.W.; Lentink, D.; Oh, P.Y.; Siegwart, R. Flying Robots. In *Springer Handbook of Robotics*; Siciliano, B.; Khatib, O., Eds.; Springer International Publishing: Cham, Switzerland, 2016; pp. 623–670.
3. Fahimi, F. Autonomous Helicopters. In *Autonomous Robots: Modeling, Path Planning, and Control*; Springer: Boston, MA, USA, 2009; pp. 263–317.
4. Gerdes, J.; Holness, A.; Perez-Rosado, A.; Roberts, L.; Greisinger, A.; Barnett, E.; Kempny, J.; Lingam, D.; Yeh, C.H.; Bruck, H.A.; et al. Robo Raven: A Flapping-Wing Air Vehicle with Highly Compliant and Independently Controlled Wings. *Soft Robot.* **2014**, *1*, 275–288. [CrossRef]

5. Burri, M.; Gasser, L.; Käch, M.; Krebs, M.; Laube, S.; Ledergerber, A.; Meier, D.; Michaud, R.; Mosimann, L.; Müri, L.; et al. Design and Control of a Spherical Omnidirectional Blimp. In Proceedings of the 2013 IEEE/RSJ International Conference on Intelligent Robots and Systems (IROS), Tokyo, Japan, 3–7 November 2013; pp. 1873–1879.

6. Luque-Vega, L.; Castillo-Toledo, B.; Loukianov, A.G. Robust block second order sliding mode control for a quadrotor. *J. Frankl. Inst.* **2012**, *349*, 719–739. [CrossRef]

7. Zhao, J.; Zhang, H.; Li, X. Active disturbance rejection switching control of quadrotor based on robust differentiator. *Syst. Sci. Control Eng.* **2020**, *8*, 605–617. [CrossRef]

8. Glida, H.E.; Abdou, L.; Chelihi, A.; Sentouh, C.; Hasseni, S.E.I. Optimal model-free backstepping control for a quadrotor helicopter. *Nonlinear Dyn.* **2020**, *100*, 3449–3468. [CrossRef]

9. Pérez-Alcocer, R.; Moreno-Valenzuela, J. A novel Lyapunov-based trajectory tracking controller for a quadrotor: Experimental analysis by using two motion tasks. *Mechatronics* **2019**, *61*, 58–68. [CrossRef]

10. Noormohammadi-Asl, A.; Esrafilian, O.; Ahangar Arzati, M.; Taghirad, H.D. System identification and H_∞-based control of quadrotor attitude. *Mech. Syst. Signal Process.* **2020**, *135*, 106358. [CrossRef]

11. Huynh, M.Q.; Zhao, W.; Xie, L. $\mathcal{L}_1$ adaptive control for quadcopter: Design and implementation. In Proceedings of the 2014 13th International Conference on Control Automation Robotics Vision (ICARCV), Singapore, 10–12 December 2014; pp. 1496–1501.

12. Jafarnejadsani, H.; Sun, D.; Lee, H.; Hovakimyan, N. Optimized L1 Adaptive Controller for Trajectory Tracking of an Indoor Quadrotor. *J. Guid. Control Dyn.* **2017**, *40*, 1415–1427. [CrossRef]

13. Erginer, B.; Altuğ, E. Design and implementation of a hybrid fuzzy logic controller for a quadrotor VTOL vehicle. *Int. J. Control. Autom. Syst.* **2012**, *10*, 61–70. [CrossRef]

14. Boudjedir, H.; Bouhali, O.; Rizoug, N. Adaptive neural network control based on neural observer for quadrotor unmanned aerial vehicle. *Adv. Robot.* **2014**, *28*, 1151–1164. [CrossRef]

15. Eskandarpour, A.; Sharf, I. A constrained error-based MPC for path following of quadrotor with stability analysis. *Nonlinear Dyn.* **2020**, *99*, 899–918. [CrossRef]

16. Liu, J. Introduction to Intelligent Control. In *Intelligent Control Design and MATLAB Simulation*; Springer: Singapore, 2018; pp. 1–5.

17. Dydek, Z.T.; Annaswamy, A.M.; Lavretsky, E. Adaptive Control of Quadrotor UAVs: A Design Trade Study With Flight Evaluations. *IEEE Trans. Control Syst. Technol.* **2013**, *21*, 1400–1406. [CrossRef]

18. Sankaranarayanan, V.N.; Roy, S. Introducing switched adaptive control for quadrotors for vertical operations. *Optim. Control Appl. Methods* **2020**, *41*, 1875–1888. [CrossRef]

19. Sankaranarayanan, V.N.; Roy, S.; Baldi, S. Aerial Transportation of Unknown Payloads: Adaptive Path Tracking for Quadrotors. In Proceedings of the 2020 IEEE/RSJ International Conference on Intelligent Robots and Systems (IROS), Las Vegas, NV, USA, 25–29 October 2020; pp. 7710–7715.

20. Tian, B.; Cui, J.; Lu, H.; Zuo, Z.; Zong, Q. Adaptive Finite-Time Attitude Tracking of Quadrotors With Experiments and Comparisons. *IEEE Trans. Ind. Electron.* **2019**, *66*, 9428–9438. [CrossRef]

21. Nguyen, N.P.; Mung, N.X.; Thanh, H.L.N.N.; Huynh, T.T.; Lam, N.T.; Hong, S.K. Adaptive Sliding Mode Control for Attitude and Altitude System of a Quadcopter UAV via Neural Network. *IEEE Access* **2021**, *9*, 40076–40085. [CrossRef]

22. Mehmood, Y.; Aslam, J.; Ullah, N.; Chowdhury, M.S.; Techato, K.; Alzaed, A.N. Adaptive Robust Trajectory Tracking Control of Multiple Quad-Rotor UAVs with Parametric Uncertainties and Disturbances. *Sensors* **2021**, *21*, 2401. [CrossRef]

23. Sun, C.; Liu, M.; Liu, C.; Feng, X.; Wu, H. An Industrial Quadrotor UAV Control Method Based on Fuzzy Adaptive Linear Active Disturbance Rejection Control. *Electronics* **2021**, *10*, 376. [CrossRef]

24. Yañez-Badillo, H.; Beltran-Carbajal, F.; Tapia-Olvera, R.; Valderrabano-Gonzalez, A.; Favela-Contreras, A.; Rosas-Caro, J.C. A Dynamic Motion Tracking Control Approach for a Quadrotor Aerial Mechanical System. *Shock Vib.* **2020**, *2020*, 6635011.

25. Spong, M.; Hutchinson, S.; Vidyasagar, M. *Robot Modeling and Control*, 2nd ed.; Wiley: Hoboken, NJ, USA, 2020.

26. Fliess, M.; Marquez, R.; Delaleau, E.; Sira-Ramirez, H. Correcteurs proportionnels-intégraux généralisés. *ESAIM Control Optim. Calc. Var.* **2002**, *7*, 23–41. [CrossRef]

27. Yao, J.; Deng, W. Active disturbance rejection adaptive control of uncertain nonlinear systems: Theory and application. *Nonlinear Dyn.* **2017**, *89*, 1611–1624. [CrossRef]

28. Yañez-Badillo, H.; Tapia-Olvera, R.; Beltran-Carbajal, F. Adaptive Neural Motion Control of a Quadrotor UAV. *Vehicles* **2020**, *2*, 468–490. [CrossRef]

29. Brown, M.; Harris, C. *Neurofuzzy Adaptive Modelling and Control*; Prentice Hall International (UK) Ltd.: Hertfordshire, UK, 1994.

30. Saad, D. (Ed.) *On-Line Learning in Neural Networks*; Cambridge University Press: New York, NY, USA, 1998.

31. Ravankar, A.; Ravankar, A.A.; Kobayashi, Y.; Hoshino, Y.; Peng, C.C. Path Smoothing Techniques in Robot Navigation: State-of-the-Art, Current and Future Challenges. *Sensors* **2018**, *18*, 3170. [CrossRef] [PubMed]

32. Beltran-Carbajal, F.; Tapia-Olvera, R.; Valderrabano-Gonzalez, A.; Yanez-Badillo, H.; Rosas-Caro, J.; Mayo-Maldonado, J. Closed-loop online harmonic vibration estimation in DC electric motor systems. *Appl. Math. Model.* **2021**, *94*, 460–481. [CrossRef]

33. Beltran-Carbajal, F.; Silva-Navarro, G. On the algebraic parameter identification of vibrating mechanical systems. *Int. J. Mech. Sci.* **2015**, *92*, 178–186. [CrossRef]

34. Conyers, S.A.; Rutherford, M.J.; Valavanis, K.P. An Empirical Evaluation of Ground Effect for Small-Scale Rotorcraft. In Proceedings of the 2018 IEEE International Conference on Robotics and Automation, Brisbane, Australia, 21–26 May 2018; pp. 1244–1250.

35. Li, D.; Zhou, Y.; Shi, Z.; Lu, G. Autonomous landing of quadrotor based on ground effect modelling. In Proceedings of the 2015 34th Chinese Control Conference, Hangzhou, China, 28–30 July 2015; pp. 5647–5652.
36. Marini, F.; Walczak, B. Particle swarm optimization (PSO). A tutorial. *Chemom. Intell. Lab. Syst.* **2015**, *149*, 153–165. [CrossRef]
37. Zhao, J.; Li, T.; Qian, J. Application of Particle Swarm Optimization Algorithm on Robust PID Controller Tuning. In *Advances in Natural Computation*; Wang, L., Chen, K., Ong, Y.S., Eds.; Springer: Berlin/Heidelberg, Germany, 2005; pp. 948–957.
38. Kaminski, M. Neural Network Training Using Particle Swarm Optimization—A Case Study. In Proceedings of the 2019 24th International Conference on Methods and Models in Automation and Robotics (MMAR), Międzyzdroje, Poland, 26–29 August 2019; pp. 115–120.
39. Shi, D.; Wu, Z.; Chou, W. Generalized Extended State Observer Based High Precision Attitude Control of Quadrotor Vehicles Subject to Wind Disturbance. *IEEE Access* **2018**, *6*, 32349–32359. [CrossRef]
40. Sreenath, K.; Michael, N.; Kumar, V. Trajectory generation and control of a quadrotor with a cable-suspended load—A differentially-flat hybrid system. In Proceedings of the 2013 IEEE International Conference on Robotics and Automation, Karlsruhe, Germany, 6–10 May 2013; pp. 4888–4895.
41. Mellinger, D.; Kumar, V. Minimum snap trajectory generation and control for quadrotors. In Proceedings of the 2011 IEEE International Conference on Robotics and Automation, Shanghai, China, 9–13 May 2011; pp. 2520–2525.
42. Beltran-Carbajal, F.; Silva-Navarro, G. A fast parametric estimation approach of signals with multiple frequency harmonics. *Electr. Power Syst. Res.* **2017**, *144*, 157–162. [CrossRef]

 *mathematics*

MDPI

Article

Active Suspension Control Using an MPC-LQR-LPV Controller with Attraction Sets and Quadratic Stability Conditions

Daniel Rodriguez-Guevara [1], Antonio Favela-Contreras [1,*], Francisco Beltran-Carbajal [2], David Sotelo [1] and Carlos Sotelo [1]

1   Tecnologico de Monterrey, School of Engineering and Sciences, Ave. Eugenio Garza Sada 2501, Monterrey 64849, Mexico; A01280937@itesm.mx (D.R.-G.); david.sotelo@tec.mx (D.S.); carlos.sotelo@tec.mx (C.S.)
2   Departamento de Energía, Universidad Autónoma Metropolitana, Unidad Azcapotzalco, Av. San Pablo No. 180, Col. Reynosa Tamaulipas, Mexico City 02200, Mexico; fbeltran.git@gmail.com
*   Correspondence: antonio.favela@tec.mx

Abstract: The control of an automotive suspension system by means of a hydraulic actuator is a complex nonlinear control problem. In this work, a Linear Parameter Varying (LPV) model is proposed to reduce the complexity of the system while preserving the nonlinear behavior. In terms of control, a dual controller consisting of a Model Predictive Control (MPC) and a Linear Quadratic Regulator (LQR) is implemented. To ensure stability, Quadratic Stability conditions are imposed in terms of Linear Matrix Inequalities (LMI). Simulation results for quarter-car model over several disturbances are tested in both frequency and time domain to show the effectiveness of the proposed algorithm.

Keywords: active suspension; model predictive control; linear parameter varying; ellipsoidal set; attraction sets; quadratic stability

Citation: Rodriguez-Guevara, D.; Favela-Contreras, A.; Beltran-Carbajal, F.; Sotelo, D.; Sotelo, C. Active Suspension Control Using an MPC-LQR-LPV Controller with Attraction Sets and Quadratic Stability Conditions. *Mathematics* **2021**, *9*, 2533. https://doi.org/10.3390/math9202533

Academic Editor: Paolo Mercorelli

Received: 26 July 2021
Accepted: 25 August 2021
Published: 9 October 2021

Publisher's Note: MDPI stays neutral with regard to jurisdictional claims in published maps and institutional affiliations.

1. Introduction

A vehicle can experiment different road disturbances while maneuvering in normal conditions such as bumps or bends. The suspension system of a car is designed to attenuate those disturbances to preserve comfort for the passengers while maintaining safe driving conditions to control the car's direction. However, when road conditions are harsh, passive suspension systems may fail to preserve both comfort and road holding.

The Active Suspension system has been used to improve road-holding conditions while improving the comfort of passengers by means of a hydraulic actuator. To provide Active Suspension control, several control strategies have been proposed in the literature such as PID controller [1–4], H2 and H∞ control [5–9], fuzzy logic control [10–13], and sliding mode control [14–17]. All these controllers have exhibited a trade-off between comfort and road holding, with specific tuning conditions to manage each one of the design specifications according to the desired performance.

Another control strategy widely used in Active Suspensions is Model Predictive Control (MPC). MPC approaches encompass several MIMO control strategies involving the prediction of the future behavior of the system along a prediction horizon N_p and finding an optimal control solution subject to constraints in inputs, outputs and states. The general structure of the MPC strategy is to solve at each step an optimization problem where a cost function is minimized subject to constraints to find the optimal input sequence to be introduced in the system.

Some MPC approaches for Active Suspension systems are the following. In [18] an autoregressive with exogenous variable (ARX) model-based predictive control is presented to improve passenger comfort and road holding in a vehicle using a semi-Active Suspension with a Bouc–Wen representation. The results showed improvement when compared with passive suspension; however, the results were limited by the type of suspension used in this application. In [19] a full-car suspension model is controlled by a linear MPC with 6 degrees of freedom (6-DOF). In this approach, the control goal is to achieve a desired tilt angle to preserve comfort and road holding. The actuator is considered to be ideal, which results in a linear behavior of the system. This allows real-time implementation due to the short optimization time; however, the performance of the control algorithm may not be as effective as the one shown in simulation due to the linear design of the controller. Other MPC approaches using linear models are presented in [20–22].

Another MPC approach for the Active Suspension system is presented in [23]. In previous work, the Active Suspension quarter-car system was modeled as nonlinear by considering the nonlinear effect of the actuator. To comply with the MPC approach, the system is modeled as a Takagi–Sugeno model (T–S) by a fuzzy representation consisting of two sub-models. The Model Predictive Control is also designed as a fuzzy MPC where there exist two interconnected linear models and the switching between one model and the another is performed by fuzzy logic. Additionally, terminal equality constraints are included in terms of Linear Matrix Inequalities (LMI) to ensure stability. In [24] a robust model predictive controller (RMPC) for an Active Suspension full-car system is presented. This approach considers both the nonlinearities of the hydraulic actuator and the nonlinearities presented by the relationship of the movement of each one of the four corners of the car. Therefore, to design the MPC, the model is simplified into a linear fuzzy logic system. To add robustness to the control approach, an adaptive control law is proposed based on the MPC and a fractional PID controller.

As shown in the previous works, the nonlinear model of an Active Suspension system is represented by linear representations to comply with the MPC strategy. In this research work, a Linear Parameter Varying (LPV) representation is proposed. This kind of representation is common in semi-Active Suspension control approaches [25–27]. In this approach, the LPV representation is done using one scheduling parameter embedding the nonlinearities of the hydraulic actuator.

Therefore, the proposed control strategy consists of a Model Predictive Controller for an Active Suspension system with an electro-hydraulic actuator with a servo spool valve. The model of the system is constructed as a Linear Parameter Varying model using one scheduling variable ρ_1. Quadratic Stability conditions are included in the MPC algorithm as LMI, as presented in [28]. To improve performance, a terminal cost using attraction sets is included, as shown in [29]. Finally, the inclusion of a terminal set and a Linear Quadratic Regulator (LQR) controller in the terminal set is included.

The rest of the paper is organized in the following structure. Section 2 presents the Active Suspension with electro-hydraulic actuator model. Section 3 shows a state–space LPV representation of the Active Suspension. Section 4 describes the MPC-LPV control algorithm. Section 5 shows a Recursive Least Squares (RLS) algorithm for the prediction of the scheduling parameter along the prediction horizon. Section 6 presents the Quadratic Stability conditions for the MPC-LPV approach. Section 7 describes the attraction sets and terminal set for control switching. Results and simulations are presented in Section 8 and conclusions are discussed in Section 9.

2. Quarter-Car Active Suspension Model

Active Suspension systems add an actuator to the passive system mainly consisting of the wheel mass and the chassis mass. Figure 1 presents a schematic model of an Active Suspension system as found in [7]. In this model, the actuator produces a force f_s which reduces the vertical movement of both masses m_s, which is the sprung mass representing the chassis body, and m_{us}, which is the unsprung mass representing the suspension unit and wheel of the quarter-car.

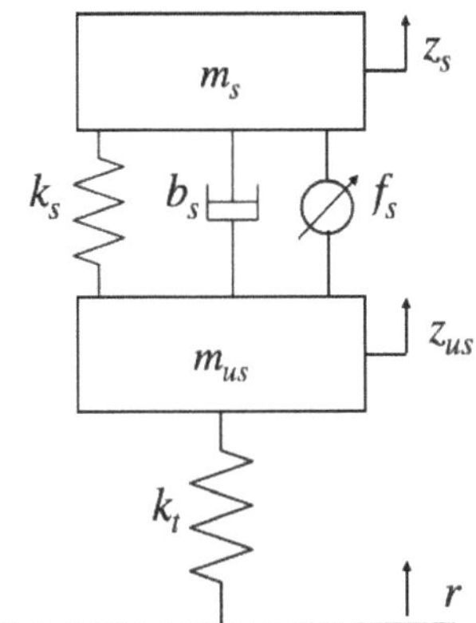

Figure 1. Active Suspension System.

The dynamic force equations of the system are the following:

$$m_s\ddot{z}_s + k_s(z_s - z_{us}) + b_s(\dot{z}_s - \dot{z}_{us}) - f_s = 0 \tag{1}$$

$$m_{us}\ddot{z}_{us} - k_s(z_s - z_{us}) - b_s(\dot{z}_s - \dot{z}_{us}) + k_t\left(z_{us} - r(t)\right) + f_s = 0 \tag{2}$$

with k_s being the constant of the spring between the two masses and b_s being the damping coefficient. k_t represents the tire elastic constant and $r(t)$ represents the road disturbances. The force f_s is generated by an electro-hydraulic actuator with a servo spool valve. A schematic of the electro-hydraulic actuator is shown in Figure 2. By means of this actuator, the force f_s is generated by the pressure supplied to the system P_l and the area of the moving piston A. Therefore, the force can be expressed using the following equation.

$$f_s = AP_l \tag{3}$$

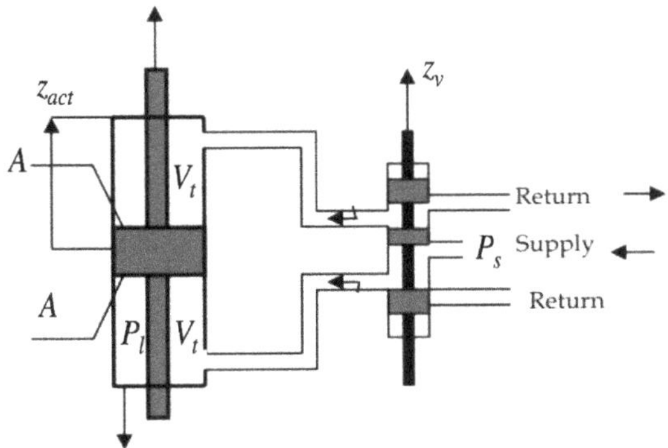

Figure 2. Electro-Hydraulic Actuator.

The pressure P_l dynamics are modeled using the following differential Equation (4).

$$\frac{V_t}{4\beta_e}\dot{P}_l = Q - C_{tp}P_l - A(\dot{z}_s - \dot{z}_{us}) \tag{4}$$

with $Q = \mathrm{sgn}\left[P_s - \mathrm{sgn}(z_v)P_l\right]C_d w x_v \sqrt{\frac{1}{\rho}\left|P_s - \mathrm{sgn}(z_v)P_l\right|}$, where V_t is the total actuator volume, Q the load flow, B_e the effective bulk modulus, C_{tp} the piston leakage coefficient, C_d the discharge coefficient, w the spool valve area gradient, ρ the hydraulic fluid density and P_s the pressure supply. The displacement of the spool valve z_v is proportional to the control action $u(k)$ which is a voltage signal. The valve displacement is expressed by the following equation.

$$\dot{z}_v = \frac{z_v}{\tau} + k_v u(k) \tag{5}$$

where k_v represents the valve gain and τ is a scaling factor.

3. LPV-SS Representation of the Quarter-Car Active Suspension Model

To comply with the MPC strategy, Equations (1) through (5) will be used to build a Linear Parameter Varying state–space (LPV-SS) model with a scheduling parameter ρ_1 of the form:

$$\dot{x}(t) = \mathbf{A}\left(\rho_1(t)\right)x(t) + \mathbf{B}u(t) + \mathbf{B_r}r(t) \tag{6}$$

with the discrete LPV-SS obtained by a Zero-Order Hold (ZOH) represented as:

$$x(k+1) = \mathbf{A}\left(\rho_1(k)\right)x(k) + \mathbf{B}u(k) + \mathbf{B_r}r(k) \tag{7}$$

where $\dot{\mathbf{x}}(t) = [x_1\ x_2\ x_3\ x_4\ x_5\ x_6]^T = [z_s\ \dot{z}_s\ z_{us}\ \dot{z}_{us}\ P_l\ z_v]^T$. With $\mathbf{A}$ being the state matrix and $\mathbf{B}$ the input matrix, $\mathbf{B_r}$ represents the input disturbance matrix while $u(k)$ is the control input. Therefore, the space-state matrices can be defined as the following:

$$\mathbf{A}\left(\rho_1(k)\right) = \begin{bmatrix} 0 & 1 & 0 & 0 & 0 & 0 \\ -\frac{k_s}{m_s} & -\frac{b_s}{m_s} & \frac{k_s}{m_s} & \frac{b_s}{m_s} & A & 0 \\ 0 & 0 & 0 & 1 & 0 & 0 \\ \frac{k_s}{m_{us}} & \frac{b_s}{m_{us}} & -\frac{k_s+k_t}{m_{us}} & -\frac{b_s}{m_{us}} & -\frac{A}{m_{us}} & 0 \\ 0 & -\alpha A & 0 & \alpha A & -\beta & \gamma\rho_1 \\ 0 & 0 & 0 & 0 & 0 & \frac{1}{\tau} \end{bmatrix} \tag{8}$$

$$\mathbf{B} = \begin{bmatrix} 0 \\ 0 \\ 0 \\ 0 \\ 0 \\ \frac{k_v}{\tau} \end{bmatrix} \tag{9}$$

$$\mathbf{B_r} = \begin{bmatrix} 0 \\ 0 \\ 0 \\ \frac{k_t}{m_{us}} \\ 0 \\ 0 \end{bmatrix} \tag{10}$$

With $\alpha = \frac{4\beta_e}{V_t}$, $\beta = \alpha C_{tp}$, $\gamma = \alpha C_d \sqrt{\frac{1}{\rho}}$ and $\rho_1 = \mathrm{sgn}\left[P_s - \mathrm{sgn}(x_6)x_5\right]\sqrt{\left|P_s - \mathrm{sgn}(x_6)x_5\right|}$. The inclusion of the scheduling variable ρ_1 allows the system to be expressed as an LPV-SS representation which allows the MPC law to be computed in a compact matrix form.

4. LPV-MPC Controller

To develop a MPC scheme for LPV models, the future states need to be formulated so a trajectory can be formed along the prediction horizon. The *i*-steps-ahead prediction can be structured as the following:

$$\mathbf{x}(k+i|k) = \left(\prod_{j=0}^{i-1} \mathbf{A}\Big(\rho_1(k+j)\Big)\right)\mathbf{x}(k) + \left(\sum_{s=1}^{i-1}\left(\prod_{l=s}^{i-1}\mathbf{A}\Big(\rho_1(k+l)\Big)\right)\right)\mathbf{B}u(k+s-1)\right) + \mathbf{B}u(k+i-1) \tag{11}$$

Prediction of the future states needs to be performed for the future N_p time steps, thus using (11) the following matrix equation can be deduced.

$$\mathbf{X} = \mathbf{\Phi} * \mathbf{x}(k) + \mathbf{\Psi} * \mathbf{U} \tag{12}$$

Where:

$$\mathbf{X} = \begin{bmatrix} x(k+1|k) \\ x(k+2|k) \\ \vdots \\ x(k+N_p|k) \end{bmatrix} \tag{13}$$

$$\mathbf{\Phi} = \begin{bmatrix} \mathbf{A}(\rho_1(k)) \\ \prod_{j=0}^{1}\Big(\mathbf{A}\Big(\rho_1(k+j)\Big)\Big) \\ \vdots \\ \prod_{j=0}^{N_p-1}\Big(\mathbf{A}\Big(\rho_1(k+j)\Big)\Big) \end{bmatrix} \tag{14}$$

$$\mathbf{\Psi} = \begin{bmatrix} \mathbf{B} & \mathbf{0}_{n_x \times n_u} & \cdots & \mathbf{0}_{n_x \times n_u} \\ \mathbf{A}\Big(\rho_1(k+1)\Big)\mathbf{B} & \mathbf{B} & \cdots & \mathbf{0}_{n_x \times n_u} \\ \mathbf{A}\Big(\rho_1(k+2)\Big)\mathbf{A}\Big(\rho_1(k+1)\Big)\mathbf{B} & \mathbf{A}\Big(\rho_1(k+2)\Big)\mathbf{B} & \cdots & \mathbf{0}_{n_x \times n_u} \\ \vdots & \vdots & \ddots & \vdots \\ \left(\prod_{i=1}^{N_p-1}\mathbf{A}\Big(\rho_1(k+i)\Big)\right)\mathbf{B} & \left(\prod_{i=1}^{N_p-1}\mathbf{A}\Big(\rho_1(k+i+1)\Big)\right)\mathbf{B} & \cdots & \mathbf{B} \end{bmatrix} \tag{15}$$

$$\mathbf{U} = \begin{bmatrix} u(k) \\ u(k+1) \\ \vdots \\ u(k+N_p-1) \end{bmatrix} \tag{16}$$

with $\mathbf{X} \in \mathbb{R}^{N_p \cdot n_x}$, $\mathbf{\Phi} \in \mathbb{R}^{N_p \cdot n_x \times n_x}$, $\mathbf{\Psi} \in \mathbb{R}^{N_p \cdot n_x \times N_p \cdot n_u}$ and $\mathbf{U} \in \mathbb{R}^{N_p \cdot n_u}$ where n_x is the number of states and n_u the number of inputs. With the state prediction equation, we can construct a cost function to minimize the deviation from the equilibrium states and the energy used by the inputs, so that the cost function is defined as:

$$J = \mathbf{X}^T \mathbf{Q_c} \mathbf{X} + \mathbf{U}^T \mathbf{R_c} \mathbf{U} \tag{17}$$

where $\mathbf{Q_c}$ and $\mathbf{R_c}$ are weight matrix of appropriate dimensions. To find the optimal control trajectory $\mathbf{U}$, (17) needs to be minimized subject to the constraints in the inputs (18) and the states (19).

$$u_{min} \leq \mathbf{U} \leq u_{max} \tag{18}$$

$$x_{min} \leq \mathbf{X} \leq x_{max} \tag{19}$$

with both u_{min} & $u_{max} \in \mathbb{R}^{N_p \times n_u}$ and both x_{min} & $x_{max} \in \mathbb{R}^{N_p \times N_p}$. However, to properly solve the MPC problem, the future values of ρ_1, which are unknown, must be estimated. To obtain the values we must estimate them using an RLS approach to obtain an approximate value of the scheduling parameter based on its previous behavior and the system response.

5. Scheduling Parameter Prediction Using RLS

To obtain an estimation of the future scheduling parameter, an RLS approach is used as presented by Sename, Morato & Normey-Rico in [26]. The scheduling parameter is assumed to be measurable at instant k and all previous values can be stored; however, the future parameters will be estimated based on the previous measurements of the scheduling parameters as well as the previous inputs and outputs to consider the behavior of the system.

The behavior of the scheduling parameter of a LPV system can be approximated by a linear ARX model, which is a function of the previous scheduling parameter values, the previous inputs and the previous outputs. This ARX model can be represented as:

$$\rho_1(k + N_p) = a_0\rho_1(k) + ... + a_{N_p}\rho_1(k - N_p) + b_0 u(k - 1) + ...$$

$$... + b_{N_p}u(k - N_p - 1) + c_0 y(k) + ... + c_{N_p}y(k - Np) \tag{20}$$

Afterwards, (20) can be expressed in a compact form and be dependent only on known values to be suitable for MPC design. To find a solution to the RLS, parameters a_0 to c_{N_p} need to be calculated. These parameters will be grouped into the following vector:

$$\Theta(k) = [a_0...c_{N_p}]^T \tag{21}$$

resulting in:

$$\rho_1(k) = \gamma(k)^T\Theta(k) \tag{22}$$

with:

$$\gamma(k)^T = \left[\rho_1(k - N_p), ..., \rho_1(k - 2N_p), u(k - N_p - 1), u(k - 2N_p - 1), y(k - N_p), ..., y(k - 2N_p)\right] \tag{23}$$

with (22) and (23) a direct solution can be built and used to find ρ_1 in an online RLS algorithm as presented in [30]:

$$\Theta(k) = \Theta(k - 1) + \sigma(k)\left(\rho_1(k) - \gamma(k - 1)^T\Theta(k - 1)\right) \tag{24}$$

$$\hat{Q}(k) = \left(I - \sigma(k)\gamma(k)^T\right)\frac{\hat{Q}(k - 1)}{\mu} \tag{25}$$

with $\mu \in [0, 1]$ being a forgetting factor that gives exponentially less weight to older error samples of the RLS algorithm and $\sigma(k)$ being a vector defined as:

$$\sigma(k) = \frac{1}{\mu c(k)}\hat{Q}(k - 1)\gamma(k) \tag{26}$$

and $c(k)$ is a scalar defined by:

$$c(k) = 1 + \gamma(k)^T\frac{\hat{Q}(k - 1)}{\mu}\gamma(k) \tag{27}$$

Therefore, the RLS algorithm for estimating the future scheduling parameters is shown as Algorithm 1:

After solving the RLS algorithm for the N_p future scheduling parameters, they will be considered to be known and exact to build a vector $\hat{P}(k) = \left[\rho_1(k), ..., \rho_1(k + N_p)\right]^T$ which contains all of them; therefore, (12) is no longer an equation with unknown variables and can be solved through LMI optimization.

Algorithm 1

Offline
Step 1—Initialize $\boldsymbol{\Theta}(0)$ and $\hat{\mathbf{Q}}(0)$
Online
Step 2—Obtain $\rho_1(k)$, $y(k)$ and $u(k)$
Step 3—Construct $\gamma^T(k)$ vector
Step 4—Calculate scalar c
Step 5—Obtain vector $\sigma(k)$
Step 6—Obtain $\boldsymbol{\Theta}(k)$
Step 7—Obtain $\hat{\mathbf{Q}}(k)$
Step 8—Obtain $\rho_1(k)$
Step 9—Set $k = k + 1$, If $k < N_p$ go to Step 10, else, go back to step 3
Step 10—Construct $\hat{\mathbf{P}}(k) = \left[\rho_1(k), \rho_1(k+1), ..., \rho_1(k+N_p) \right]$

---

6. Quadratic Stability in MPC-LPV Approach

To ensure Quadratic Stability in the MPC-LPV approach, system (7) can be considered to be a parametric uncertain system. In parametric uncertain systems, the scheduling variable is limited to vary in a range $\Delta\rho_{1min} \leq \Delta\rho_{1k} \leq \Delta\rho_{1max}$. To ensure stability in parametric uncertain systems, the following condition needs to be met as presented in [31].

$$\left(\mathbf{A}(\rho_1) + \mathbf{BK}\right)^T \mathbf{P}\left(\mathbf{A}(\rho_1) + \mathbf{BK}\right) - \mathbf{P} < 0 \tag{28}$$

which is the Riccati Equation for parametric uncertain systems where $\mathbf{P} > 0$ is a positive definite matrix of appropriate dimensions and $\mathbf{K}$ a static feedback gain matrix. Then, (28) can be pre- and post-multiplied by a matrix $\mathbf{Q} = \mathbf{P}^{-1}$ and $\mathbf{KQ} = \mathbf{R}$ to obtain:

$$\left(\mathbf{QA}^T(\rho_1) + \mathbf{R}^T\mathbf{B}^T\right)\mathbf{Q}^{-1}\left(\mathbf{A}(\rho_1)\mathbf{Q} + \mathbf{BR}\right) - \mathbf{Q} < 0 \tag{29}$$

To cope with the MPC paradigm, the Schur complement is applied to (29) to obtain the following LMI:

$$\begin{bmatrix} \mathbf{Q} & \mathbf{QA}^T(\rho_1) + \mathbf{RB} \\ \mathbf{A}(\rho_1)\mathbf{Q} + \mathbf{BR} & \mathbf{Q} \end{bmatrix} > 0 \tag{30}$$

for every possible value of ρ_1 at time instant k which leads to an infinite number of LMI. However, as system (7) is considered to be a parametric uncertain system, (30) can be evaluated on the vertex of matrix $\mathbf{A}$ to consider the worst-case scenarios. Therefore, (30) can be written as:

$$\begin{bmatrix} \mathbf{Q} & \mathbf{QA}_{i,j}^T(\rho_1) + \mathbf{RB} \\ \mathbf{A}_{i,j}(\rho_1)\mathbf{Q} + \mathbf{BR} & \mathbf{Q} \end{bmatrix} > 0 \tag{31}$$

The previous condition must be met $\forall j \in [k, k + N_p]$ and $\forall i \in [1, 2^l]$, where l is the number of scheduling variables ρ_1, $\mathbf{Q} > 0$ is a positive definite stability matrix to be determined, and $\mathbf{KQ} = \mathbf{R}$, where $\mathbf{K}$ is the static feedback gain matrix. With these adjustments, the number of LMI to be solved is now finite and equal to $2^l N_p$. Since there is the consideration of a static feedback gain, the control law is determined as $u(k) = \mathbf{Kx}(k)$, but to comply with the MPC paradigm, the previous expression can be considered to be an inequality as $u(k) < \mathbf{Kx}(k)$. This leads to a conservative MPC performance due to the limitations of the input variable. However, this problem will be addressed in Section 7 with the inclusion of terminal sets. Therefore, using (17)–(19) and (31) the optimization problem needs to find the optimal control sequence at each time step k is the following:

$$\min_{\mathbf{U}} J \ \text{ s.t.}(18), (19) \ \& \ (31) \tag{32}$$

7. MPC-LQR for LPV Models

7.1. Attraction Sets and Terminal Set

The inclusion of LMIs to ensure robust stability to the MPC paradigm often leads to a conservative performance of the control of the system. Therefore, to steer the system into a desired equilibrium state in the presence of disturbance or uncertainty, a series of terminal sets can be defined. In [32] a set of shrinking ellipsoids is determined using a decay rate, which can vary the speed of the system and the stability determined by similar stability conditions to the ones shown in Section 6, to steer the states to the equilibrium point. However, the determination of the decay rate and the constructions of the ellipsoids make this algorithm too slow for real-time applications and is rather implemented as an offline algorithm. In [33] a set of ellipsoidal sets are defined to predict the behavior of the system in the presence of bounded disturbances and uncertain bounded parameter changes. In [34] a path of ellipsoids is defined to predict the possible behavior of the scheduling parameter along the prediction horizon. In all three approaches, the goal of the ellipsoidal sets is that the states reach a terminal set or a terminal point, where a state-dependent stationary gain is applied to the system instead of the MPC law.

In this work, the future scheduling parameter is not known but predicted using the RLS algorithm presented in Section 3; therefore, the ellipsoids to build do not consider a variation on the scheduling parameter but rather the prediction error generated by the RLS algorithm. To generate the optimal desired trajectory to the setpoint, a path must be defined from every possible initial state to the terminal ellipsoidal set.

To steer the system into the desired terminal set, a term J_{TS} is added to the cost function J presented in (17). J_{TS} is defined as the following:

$$J_{TS} = \left(\mathbf{x}(k + N_p) - (\mathbf{x}_{ds} + \mathbf{x}_{dist}) \right)^T \mathbf{L} \left(\mathbf{x}(k + N_p) - (\mathbf{x}_{ds} + \mathbf{x}_{dist}) \right) - E(\rho_1) \tag{33}$$

where $E(\rho_1) = \sum_{i=k}^{k+N_p} \left[\rho_1(i) - \gamma(i-1)^T \Theta(i-1) \right]^2$ represents the sum of the squared errors of the prediction of the future parameter values. $\mathbf{x}(k + N_p)$ are the predicted states at the end of the prediction horizon, $\mathbf{x}_{ds}$ represents the desired state after N_p steps, $\mathbf{x}_{dist}$ is the predicted effect of the disturbance on the states N_p steps ahead and it was obtained by performing an open loop simulation of every possible disturbance from every initial set of states. Both $\mathbf{x}_{ds}$ and $\mathbf{x}_{dist}$ were computed offline and stored in a lookup table. $\mathbf{L}$ is a weighing matrix of appropriate dimensions. Therefore, (17) is redefined as:

$$J = \mathbf{X}^T \mathbf{Q_c} \mathbf{X} + \mathbf{U}^T \mathbf{R_c} \mathbf{U} + J_{TS} \tag{34}$$

However, the computation of every desired trajectory for every state needs to be computed offline and stored in a lookup table before the implementation of the MPC algorithm to increase execution speed.

7.2. MPC-LQR Dual Controller

To reduce the computational load of the algorithm, when the current states reach a terminal invariant set around the equilibrium point, the MPC algorithm does not need to be computed. Instead, an LQR gain can be computed based on the value of the actual prediction parameter to cope with the small error that may be present inside the terminal invariant set. The control law is then presented as:

$$u(k) = \begin{cases} \mathbf{U_{mpc}} & \mathbf{x}(k) \notin \mathbf{T} \\ \mathbf{K_{LQR}}(\rho_1)\mathbf{x}(k) & \mathbf{x}(k) \in \mathbf{T} \end{cases} \tag{35}$$

where $\mathbf{K_{LQR}}(\rho_1)$ is the LQR gain dependent on the scheduling parameter ρ_1 and $\mathbf{T}$ is the terminal invariant set defined around the equilibrium point of the system.

Figure 3 presents the block diagram for the proposed LPV-MPC-LQR control strategy. Additionally, the LPV-MPC-LQR algorithm is shown in the flowchart presented in Figure 4.

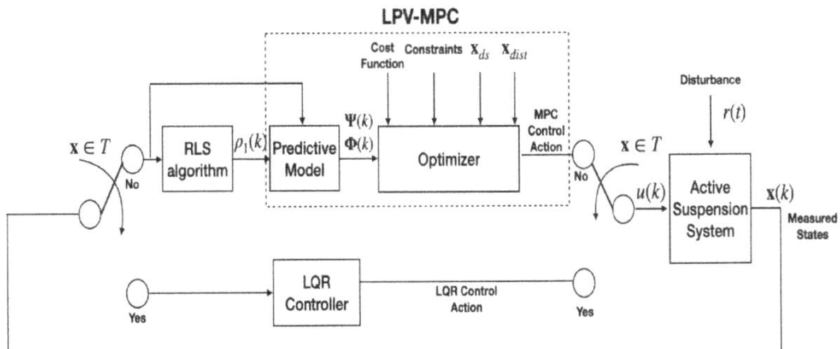

Figure 3. Block diagram of the proposed LPV-MPC-LQR control strategy for the Active Suspension system.

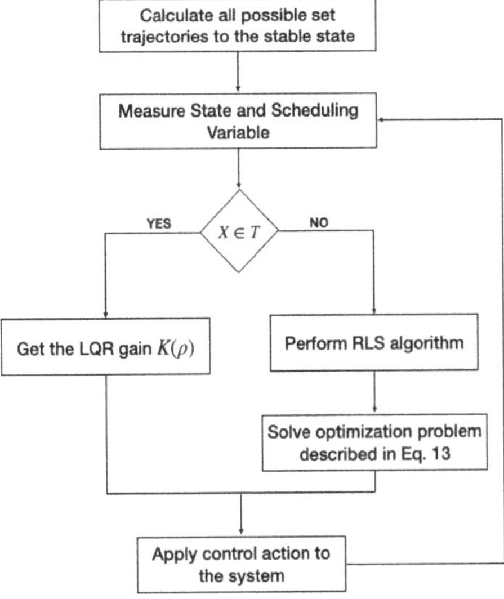

Figure 4. Flow diagram of the LPV-MPC-LQR control strategy.

8. Results and Discussion

The following simulations are made to observe the advantages and performance of implementing the proposed LQR-MPC-LPV algorithm described in Section 7. The algorithm was tested in the Active Suspension system described in Section 2. Table 1 shows the specifications of the Active Suspension system obtained from [7].

Table 1. Constant Values of the Active Suspension system.

Variable	Value	Units
m_s	250	kg
m_{us}	50	kg
m_{us}	50	kg
k_t	190,000	N/m
k_s	16,812	N/m
b_s	1000	N/(m/s)
P_s	10,300,000	Pa
τ	1/30	s
A	3.35×10^{-4}	m^2
β	1	s^{-1}
α	4.515×10^{13}	N/m^{-5}
k_v	1×10^{-4}	m/V

A discretization is made to comply with the MPC paradigm using a sampling time of $T_s = 10$ ms. A prediction horizon of $N_p = 3$ was defined after several tests using different prediction horizons were made. Using a larger prediction horizon resulted in longer optimization time and more inexact variable scheduling predictions while it does not exhibit a significant improvement in control performance. The control objective is to steer all the states to the origin while complying with the following constraints.

$$-12 \text{ V} \leq u(k) \leq 12 \text{ V}$$

$$-1 \text{ cm} \leq z_v \leq 1 \text{ cm}$$

The results will be divided into frequency-domain results and time-domain results.

8.1. Frequency-Domain Results

To obtain a frequency analysis in the nonlinear Active Suspension system, an algorithm similar to the process of defining the system response as a describing function is used. To produce these results, Algorithm 1 presented in [35] is issued. Figure 5 presents the frequency response of the Active Suspension deflection gain using the proposed LPV-MPC-LQR algorithm. Figure 6 presents the frequency response of the acceleration of the chassis mass using the proposed LPV-MPC-LQR algorithm; also, the frequency response plots are compared with the ones presented in [36] which use an LPV gain scheduling approach.

The results show how the suspension deflection is attenuated at every frequency, which results in better road holding and driving conditions. Additionally, the chassis acceleration stays in values which guarantee passenger comfort. Compared to the frequency responses of the work of Fialho et al. [36] the MPC-LQR-LPV approach presents an improvement especially in terms of road holding, shown in Figure 5, without affecting the passenger comfort.

8.2. Time-Domain Results

To obtain time-domain results using the proposed LPV-MPC-LQR control algorithm, the system was simulated using two different disturbances. Figures 7–9 present the suspension behavior when a bump disturbance of 5 cm is introduced. Figures 10–12 show the suspension behavior when driving through a sinusoidal road. The system was simulated using Matlab®; also, the software package YALMIP [37] using QP-solver SDPT3 was used for the MPC optimization. The results presented by [7] are included to make a comparison. Additionally, the results using the MPC with a frozen scheduling parameter approach without using the RLS to show the effect of the scheduling variable prediction in control performance are included.

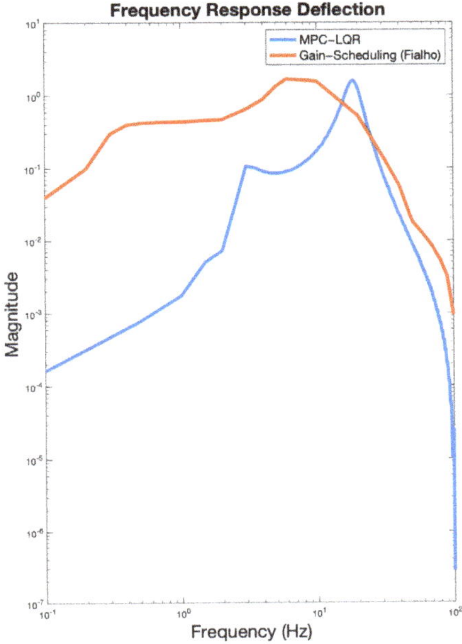

Figure 5. Frequency response of the Active Suspension deflection gain.

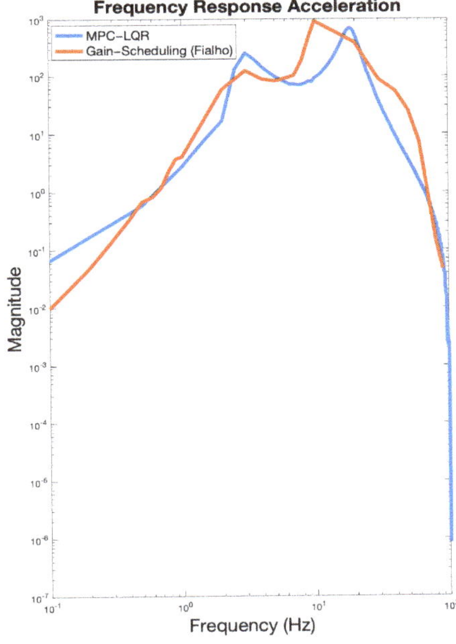

Figure 6. Frequency response of the chassis acceleration.

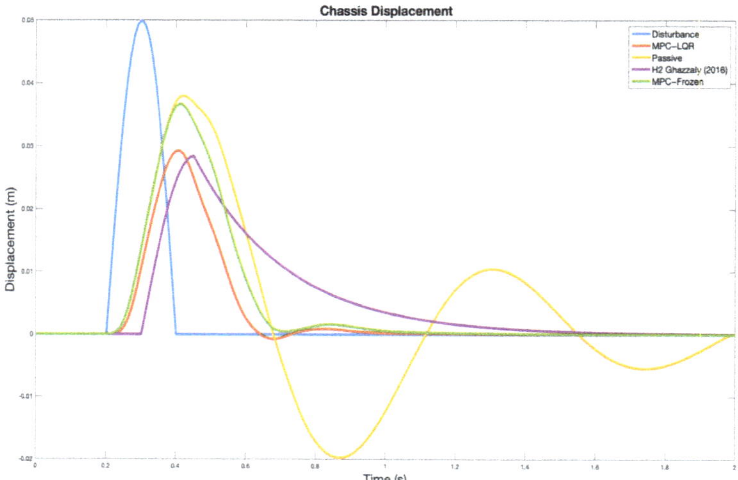

Figure 7. Chassis displacement—Bump Disturbance (Blue—disturbance, Red—MPC-LQR, Yellow—Passive, Purple—H2, Green—MPC-Frozen).

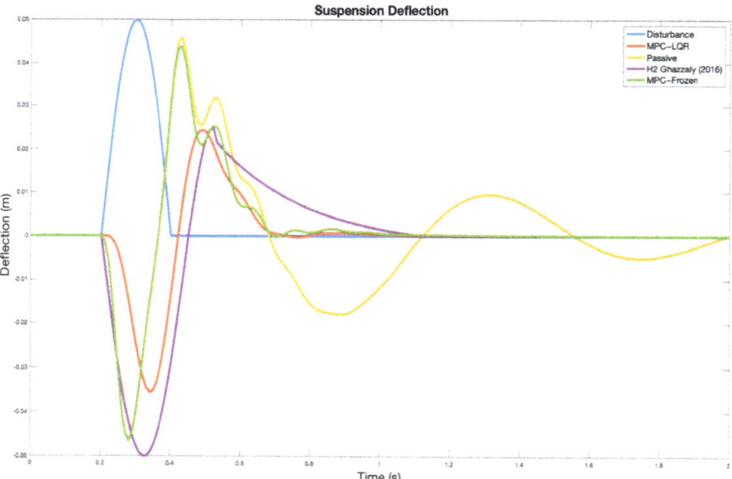

Figure 8. Suspension Deflection—Bump Disturbance (Blue—disturbance, Red—MPC-LQR, Yellow—Passive, Purple—H2, Green—MPC-Frozen).

The results of both displacement and deflection show a better performance, which results in better road holding while maintaining passenger comfort. Additionally, the comfort exhibits improvement in terms of chassis acceleration as shown in Figure 9. Additionally, to express the results numerically, both the RMS value and the maximum value of the displacement of the chassis, the suspension deflection and the acceleration of the chassis are presented in Tables 2 and 3 respectively.

Table 2. RMS Values performance.

Variable	MPC-LQR-LPV	H2 (Ghazaly, 2016)	Passive	MPC-Frozen
Chassis Displacement (m)	0.0079	0.0091	0.0142	0.0107
Suspension Deflection (m)	0.0089	0.0149	0.0240	0.0122
Chassis Acceleration (m/s^2)	0.0713	0.1104	0.1041	0.0838

Table 3. Max Values performance.

Variable	MPC-LQR-LPV	H2 (Ghazaly, 2016)	Passive	MPC-Frozen
Chassis Displacement (m)	0.0293	0.0284	0.0380	0.0367
Suspension Deflection (m)	0.0355	0.0499	0.0464	0.0439
Chassis Acceleration (m/s^2)	0.2644	0.3978	0.2925	0.2899

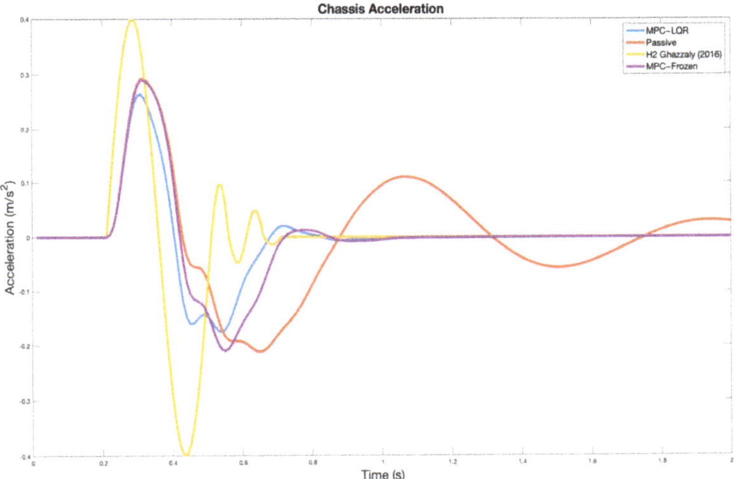

Figure 9. Chassis Acceleration—Bump disturbance (Blue—MPC-LQR, Red—Passive, Yellow—H2, Purple—MPC-Frozen).

Similar to the bump disturbance case, the proposed LPV-MPC-LQR control strategy exhibits better performance in both displacement and deflection, which results in better road holding. In terms of comfort, the acceleration of the chassis presented in Figure 12 shows a major improvement. Table 4 presents the peak values for the displacement of the chassis, the suspension deflection, and the acceleration of the chassis.

Table 4. Peak Values performance.

Variable	MPC-LQR-LPV	H2 (Ghazaly, 2016)	Passive	MPC-Frozen
Chassis Displacement (m)	0.0027	0.0044	0.0151	0.0055
Suspension Deflection (m)	0.0034	0.0040	0.0097	0.0051
Chassis Acceleration (m/s^2)	0.0164	0.32	0.162	0.0312

As shown in the previous figures, the proposed LPV-MPC-LQR control algorithm presents a better performance when compared with the H2 control strategy in both disturbance cases (bump disturbance and sinusoidal road disturbance). The RLS prediction of the future scheduling parameters have improved the control performance as well. Additionally, the proposed algorithm shows an appropriate optimization time with a worst optimization time of 930 ms and an average optimization time of 93 ms.

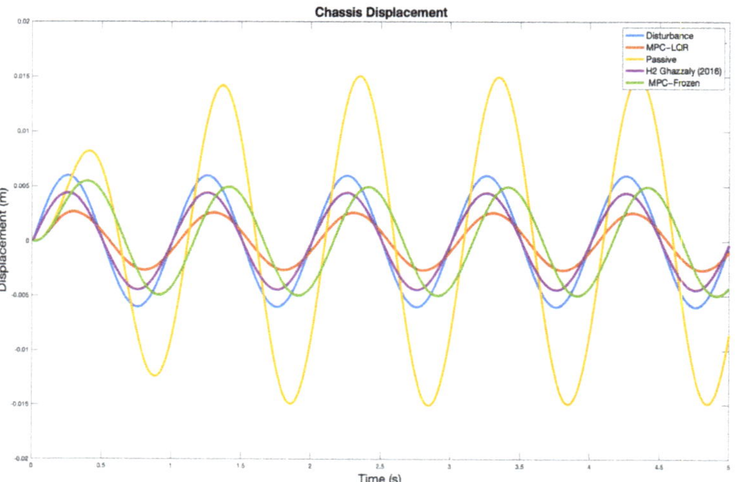

Figure 10. Chassis displacement—Sinusoidal Disturbance (Blue—disturbance, Red—MPC-LQR, Yellow—Passive, Purple—H2, Green—MPC-Frozen).

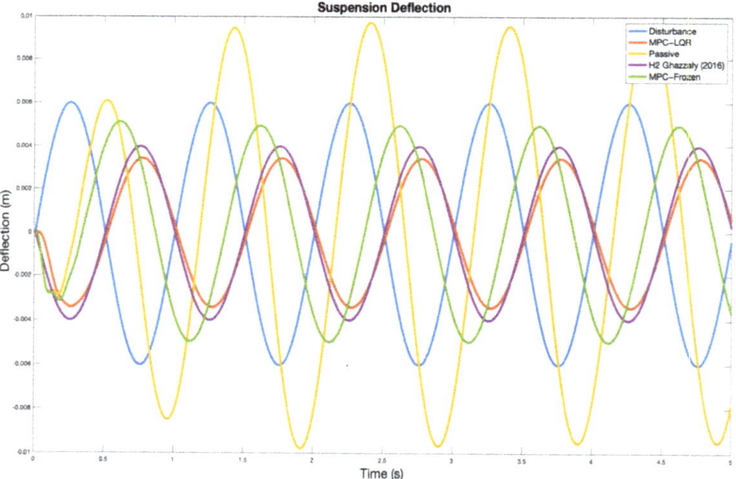

Figure 11. Suspension Deflection—Sinusoidal Disturbance (Blue—disturbance, Red—MPC-LQR, Yellow—Passive, Purple—H2, Green—MPC-Frozen).

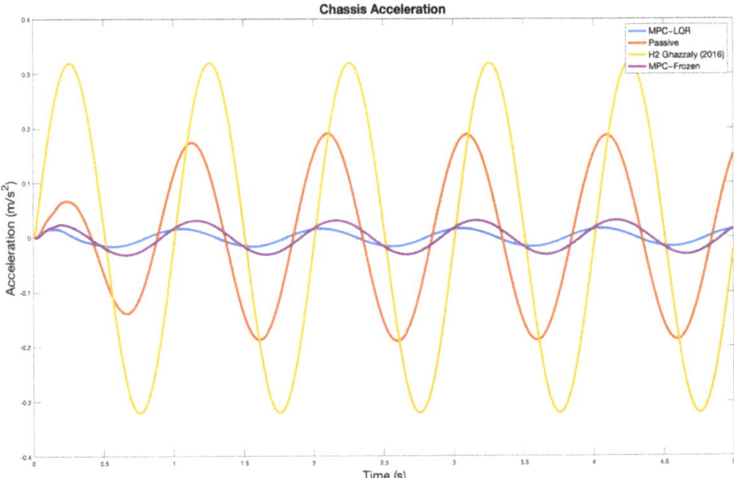

Figure 12. Chassis Acceleration—Sinusoidal disturbance (Blue—MPC-LQR, Red—Passive, Yellow—H2, Purple—MPC-Frozen).

9. Conclusions

In this paper, a novel LPV-MPC-LQR control algorithm ensuring Quadratic Stability and with the inclusion of attraction sets was presented. This method runs an RLS algorithm to obtain the prediction of the future scheduling parameter values, which simplifies the prediction of the future states while ensuring Quadratic Stability. This application can cope with nonlinear systems that can be embedded into LPV representation and therefore reduce the complexity of the algorithm and allow fast execution times. This control strategy was designed and tested on a nonlinear Active Suspension system. The results show improvements to the performance of the Active Suspension in terms of road holding and passenger comfort. Future research works should deal with recursive feasibility analysis based on stability conditions, and robustness analysis. Optimization of the LPV-MPC-LQR algorithm to achieve faster execution times using techniques of the embedded systems will also be considered in future works as well.

Author Contributions: All Authors D.R.-G., A.F.-C., F.B.-C., D.S. and C.S. have contributed as follows: Conceptualization, D.R.-G., A.F.-C., F.B.-C., D.S. and C.S.; Methodology, D.R.-G., A.F.-C., F.B.-C., D.S. and C.S.; Software, D.R.-G., D.S. and C.S.; Validation, D.R.-G., A.F.-C., F.B.-C., D.S. and C.S.; Formal analysis, D.R.-G., A.F.-C., F.B.-C., D.S. and C.S.; Investigation, D.R.-G., A.F.-C., F.B.-C., D.S. and C.S.; Writing—original draft preparation, D.R.-G. and A.F.-C.; Writing—review and editing, D.R.-G., A.F.-C., F.B.-C., D.S. and C.S.; supervision, A.F.-C. and F.B.-C.; project administration A.F.-C. All authors have read and agreed to the published version of the manuscript.

Funding: This research received no external funding.

Institutional Review Board Statement: Not applicable.

Informed Consent Statement: Not Applicable.

Data Availability Statement: No new data were created or analyzed in this study. Data sharing is not applicable to this article.

Acknowledgments: The authors would like to thank Consejo Nacional de Ciencia y Tecnología (CONACyT) and Tecnológico de Monterrey for the financial support to conduct the present research. Additionally, thanks go to the Sensors and Devices Research Group and the Robotics Research Group from the School of Engineering and Sciences of Tecnológico de Monterrey for the support given to develop this work.

Conflicts of Interest: The authors declare no conflict of interest.

References

1. Mouleeswaran, S. Design and development of PID controller-based active suspension system for automobiles. In *PID Controller Design Approaches-Theory, Tuning and Application to Frontier Areas*; Vagia, M., Ed.; Intech: Rijeka, Croatia, 2012; pp. 71–98.
2. Ahmed, A.E.N.S.; Ali, A.S.; Ghazaly, N.M.; Abd el-Jaber, G.T. PID controller of active suspension system for a quarter car model. *Int. J. Adv. Eng. Technol.* **2015**, *8*, 899.
3. Bello, M.M.; Shafie, A.A.; Khan, M.R. Electro-hydraulic pid force control for nonlinear vehicle suspension system. *Int. J. Eng. Res. Technol.* **2015**, *4*, 517–524.
4. Phu, D.X.; An, J.-H.; Choi, S.-B. A Novel Adaptive PID Controller with Application to Vibration Control of a Semi-Active Vehicle Seat Suspension. *Appl. Sci.* **2017**, *7*, 1055. [CrossRef]
5. Yu, S.; Wang, F.; Wang, J.; Chen, H. Full-car active suspension based on H2/generalised H2 output feedback control. *Int. J. Veh. Des.* **2015**, *68*, 37–54. [CrossRef]
6. Afshar, K.K.; Javadi, A.; Jahed-Motlagh, M.R. Robust H∞ control of an active suspension system with actuator time delay by predictor feedback. *IET Control. Theory Appl.* **2018**, *12*, 1012–1023. [CrossRef]
7. Ghazaly, N.M.; Ahmed, A.E.N.S.; Ali, A.S.; Abd, El-Jaber, G.T. H∞ Control of Active Suspension System for a Quarter Car Model. *Int. J. Veh. Struct. Syst. (IJVSS)* **2016**, *8*, 35–40. [CrossRef]
8. Kaleemullah, M.; Faris, W.F.; Hasbullah, F. Control of active suspension system using robust H∞ control with genetic algorithm. *Int. J. Adv. Sci. Technol.* **2019**, *28*, 763–782.
9. Jin, X.; Wang, J.; Sun, S.; Li, S.; Yang, J.; Yan, Z. Design of Constrained Robust Controller for Active Suspension of In-Wheel-Drive Electric Vehicles. *Mathematics* **2021**, *9*, 249. [CrossRef]
10. Wen, S.; Chen, M.Z.; Zeng, Z.; Yu, X.; Huang, T. Fuzzy control for uncertain vehicle active suspension systems via dynamic sliding-mode approach. *IEEE Trans. Syst. Man, Cybern. Syst.* **2016**, *47*, 24–32. [CrossRef]
11. Mustafa, G.I.; Wang, H.; Tian, Y. Model-free adaptive fuzzy logic control for a half-car active suspension system. *Stud. Inform. Control* **2019**, *28*, 13–24. [CrossRef]
12. Palanisamy, S.; Karuppan, S. Fuzzy control of active suspension system. *J. Vibroeng.* **2016**, *18*, 3197–3204. [CrossRef]
13. Alfadhli, A.; Darling, J.; Hillis, A.J. The Control of an Active Seat Suspension Using an Optimised Fuzzy Logic Controller, Based on Preview Information from a Full Vehicle Model. *Vibration* **2018**, *1*, 20–40. [CrossRef]
14. Rath, J.J.; Defoort, M.; Karimi, H.R.; Veluvolu, K.C. Output feedback active suspension control with higher order terminal sliding mode. *IEEE Trans. Ind. Electron.* **2016**, *64* 1392–1403. [CrossRef]
15. Chokor, A.; Talj, R.; Charara, A.; Shraim, H.; Francis, C. Active suspension control to improve passengers comfort and vehicle's stability. In Proceedings of the IEEE 19th International Conference on Intelligent Transportation Systems (ITSC), Rio de Janeiro, Brazil, 1–4 November 2016; pp. 296–301.
16. Taghavifar, H.; Mardani, A.; Hu, C.; Qin, Y. Adaptive robust nonlinear active suspension control using an observer-based modified sliding mode interval type-2 fuzzy neural network. *IEEE Trans. Intell. Veh.* **2019**, *5*, 53–62. [CrossRef]
17. Aljarbouh, A.; Fayaz, M. Hybrid Modelling and Sliding Mode Control of Semi-Active Suspension Systems for Both Ride Comfort and Road-Holding. *Symmetry* **2020**, *12*, 1286. [CrossRef]
18. Piñón, A.; Favela-Contreras, A.; Félix-Herrán, L.C.; Beltran-Carbajal, F.; Lozoya, C. An ARX Model-Based Predictive Control of a Semi-Active Vehicle Suspension to Improve Passenger Comfort and Road-Holding. *Actuators* **2021**, *10*, 47. [CrossRef]
19. Yao, J.; Wang, M.; Li, Z.; Jia, Y. Research on model predictive control for automobile active tilt based on active suspension. *Energies* **2021**, *14*, 671. [CrossRef]
20. Theunissen, J.; Sorniotti, A.; Gruber, P.; Fallah, S.; Ricco, M.; Kvasnica, M.; Dhaens, M. Regionless explicit model predictive control of active suspension systems with preview. *IEEE Trans. Ind. Electron.* **2019**, *67*, 4877–4888. [CrossRef]
21. Enders, E.; Burkhard, G.; Munzinger, N. Analysis of the Influence of Suspension Actuator Limitations on Ride Comfort in Passenger Cars Using Model Predictive Control. *Actuators* **2020**, *9*, 77. [CrossRef]
22. Narayan, J.; Gorji, S.A.; Ektesabi, M.M. Power reduction for an active suspension system in a quarter car model using MPC. In Proceedings of the IEEE International Conference on Energy Internet (ICEI), Sydney, Australia, 24–28 August 2020; pp. 140–146.
23. Bououden, S.; Chadli, M.; Karimi, H.R. A robust predictive control design for nonlinear active suspension systems. *Asian J. Control* **2016**, *18*, 122–132. [CrossRef]
24. Wang, D.; Zhao, D.; Gong, M.; Yang, B. Research on robust model predictive control for electro-hydraulic servo active suspension systems. *IEEE Access* **2017**, *6*, 3231–3240. [CrossRef]
25. Morato, M.M.; Sename, O.; Dugard, L. LPV-MPC fault tolerant control of automotive suspension dampers. *IFAC-PapersOnLine* **2018**, *51*, 31–36. [CrossRef]
26. Morato, M.M.; Normey-Rico, J.E.; Sename, O. Novel qLPV MPC design with least-squares scheduling prediction. *IFAC-PapersOnLine* **2019**, *52*, 158–163. [CrossRef]
27. Morato, M.M.; Normey-Rico, J.E.; Sename, O. Sub-optimal recursively feasible Linear Parameter-Varying predictive algorithm for semi-active suspension control. *IET Control Theory Appl.* **2020**, *14*, 2764–2775. [CrossRef]
28. Boyd, S.; Balakrishnan, V.; Feron, E.; ElGhaoui, L. Control system analysis and synthesis via linear matrix inequalities. In Proceedings of the American Control Conference, San Francisco, CA, USA, 2–4 June 1993; pp. 2147–2154.

29. Ferramosca, A.; Limón, D.; González, A.H.; Odloak, D.; Camacho, E.F. MPC for tracking target sets. In Proceedings of the 48h IEEE Conference on Decision and Control (CDC) Held Jointly with 2009 28th Chinese Control Conference, Shanghai, China, 15–18 December 2009; pp. 8020–8025.

30. Franklin, G.F.; Powell, J.D.; Workman, M.L. *Digital Control of Dynamic Systems*; Addison-Wesley: Reading, MA, USA, 1988; Volume 3.

31. Boyd, S.; El Ghaoui, L.; Feron, E.; Balakrishnan, V. *Linear Matrix Inequalities in System and Control Theory*; Society for Industrial and Applied Mathematics: Philadelphia, PA, USA, 1994; Chapter 5, pp. 61–76.

32. Longge, Z.; Yan, Y. Robust shrinking ellipsoid model predictive control for linear parameter varying system. *PLoS ONE* **2017**, *12*, e0178625. [CrossRef]

33. Ping, X.B.; Wang, P.; Zhang, J.F. A Multi-step Output Feedback Robust MPC Approach for LPV Systems with Bounded Parameter Changes and Disturbance. *Int. J. Control Autom. Syst.* **2018**, *16*, 2157–2168. [CrossRef]

34. Suzukia, H.; Sugie, T. MPC for LPV systems with bounded parameter variation using ellipsoidal set prediction. In Proceedings of the 2006 American Control Conference, Minneapolis, MN, USA, 14–16 June 2006; p. 6.

35. Poussot-Vassal, C.; Savaresi, S.M.; Spelta, C.; Sename, O.; Dugard, L. A methodology for optimal semi-active suspension systems performance evaluation. In Proceedings of the 49th IEEE Conference on Decision and Control (CDC), Atlanta, GA, USA, 15–17 December 2017; pp. 2892–2897.

36. Fialho, I.; Balas, G.J. Road adaptive active suspension design using linear parameter-varying gain-scheduling. *IEEE Trans. Control. Syst. Technol.* **2002**, *10*, 43–54. [CrossRef]

37. Lofberg, J. Automatic robust convex programming. *Optim. Methods Softw.* **2012**, *27*, 115–129. [CrossRef]

mathematics

Article

An Algebraic Approach for Identification of Rotordynamic Parameters in Bearings with Linearized Force Coefficients

José Gabriel Mendoza-Larios [1], Eduardo Barredo [1], Manuel Arias-Montiel [2,*], Luis Alberto Baltazar-Tadeo [3], Saulo Jesús Landa-Damas [3], Ricardo Tapia-Herrera [4] and Jorge Colín-Ocampo [3]

[1] Institute of Industrial and Automotive Engineering, Technological University of the Mixteca, Huajuapan de León 69000, Oaxaca, Mexico; jgml@mixteco.utm.mx (J.G.M.-L.); eduardin@mixteco.utm.mx (E.B.)

[2] Institute of Electronics and Mechatronics, Technological University of the Mixteca, Huajuapan de León 69000, Oaxaca, Mexico

[3] Department of Mechanical Engineering, National Technological of Mexico—CENIDET, Cuernavaca 62490, Morelos, Mexico; luis_atadeo@cenidet.edu.mx (L.A.B.-T.); saulojesuslanda@cenidet.edu.mx (S.J.L.-D.); jorge.co@cenidet.tecnm.mx (J.C.-O.)

[4] CONACYT—Technological University of the Mixteca, Huajuapan de León 69000, Oaxaca, Mexico; rtapiah@conacyt.mx

* Correspondence: mam@mixteco.utm.mx; Tel.: +52-95-3532-0214

Citation: Mendoza-Larios, J.G.;
Barredo, E.; Arias-Montiel, M.;
Baltazar-Tadeo, L.A.; Landa-Damas,
S.J.; Tapia-Herrera, R.; Colín-Ocampo,
J. An Algebraic Approach for
Identification of Rotordynamic
Parameters in Bearings with
Linearized Force Coefficients.
Mathematics **2021**, *9*, 2747. https://
doi.org/10.3390/math9212747

Academic Editor: Carlo Bianca

Received: 15 September 2021
Accepted: 25 October 2021
Published: 29 October 2021

Publisher's Note: MDPI stays neutral
with regard to jurisdictional claims in
published maps and institutional affiliations.

Abstract: In this work, a novel methodology for the identification of stiffness and damping rotordynamic coefficients in a rotor-bearing system is proposed. The mathematical model for the identification process is based on the algebraic identification technique applied to a finite element (FE) model of a rotor-bearing system with multiple degree-of-freedom (DOF). This model considers the effects of rotational inertia, gyroscopic moments, shear deformations, external damping and linear forces attributable to stiffness and damping parameters of the supports. The proposed identifier only requires the system's vibration response as input data. The performance of the proposed identifier is evaluated and analyzed for both schemes, constant and variable rotational speed of the rotor-bearing system, and numerical results are obtained. In the presented results, it can be observed that the proposed identifier accurately determines the stiffness and damping parameters of the bearings in less than 0.06 s. Moreover, the identification procedure rapidly converges to the estimated values in both tested conditions, constant and variable rotational speed.

Keywords: algebraic identification; rotor-bearing system; finite element model; rotordynamic coefficients

1. Introduction

Over the past few decades, several numerical approximations on the dynamic behavior analysis for rotordynamic systems have been developed. Among these approximations, the most popular approach is the finite element (FE) method because it is highly efficient and convenient for modelling diverse physical systems. According to Koutromanos [1], with this method a complex region that defines a continuous system is discretized with simple geometrical forms called finite elements. The material properties as well as the governing relationships are taken into consideration for these elements and expressed in terms of unknown values on the element boundaries. After an assembly process and consideration of the loads and boundary conditions, an equation system is obtained. The solution for these equations provides the approximated behavior of the continuous system. At the start of the 1960s, engineers used the FE method to obtain approximated solutions for problems related to stress analysis, fluid flows, heat transfers and other areas. However, the FE method was not applied to rotordynamics until a decade later. Through the 1970s, diverse efforts were made to incorporate effects of rotational inertia, gyroscopic moments, axial load, shear deformation and internal damping, as pointed out in [2]. Recently, Shen et al. [3] remarked on the importance of including the effects of rotational inertia in the finite elements used to model and analyze rotordynamic systems, in order to have a more general and appropriate kinematic and dynamic description of rotating structures supported by bearings with stiffness and viscous damping characteristics.

Through the bearing characterization in rotor-bearing systems, the rotordynamic stiffness and damping coefficients can be determined. Physical insight of these parameters is essential for the correct modelling of every rotordynamic system as they are important factors in determining the system's dynamical behavior. In general, when a rotor-bearing system is studied, stiffness and damping coefficients of the bearings are unknown, meaning it is therefore necessary to implement a methodology to determinate them. According to Tiwari [4], Matsushita et al. [5] and Breńkacz [6], there are eight rotordynamic coefficients in bearings, four for stiffness (two directs and two crossed) and four for damping (two directs and two crossed). Nowadays, rotor-bearing systems can be modelled in a very precise way by using modern modelling techniques. However, accurately estimating the dynamic parameters through theoretical models is still a challenge because it is difficult to accurately model every phenomenon affecting the dynamic behavior of the bearings. This problem has led to the development of novel numerical and experimental techniques for dynamic parameter identification [4,6,7]. Tiwari and Chougale [8] developed an algorithm to estimate the dynamic parameters of active magnetic bearings as well as the residual rotor unbalance. The proposed algorithm is based on the least squares technique in frequency domain. Moreover, Xu et al. [9] presented a novel identification approach for estimating bearing dynamic parameters based on the transfer matrix method. Stiffness and damping parameters of an active magnetic bearing were determined by minimizing the error between the unbalance response calculated by the transfer matrix approach and the experimental approach. Mao et al. [10] also proposed a method for identifying bearing dynamic parameters in flexible rotor-bearing systems by minimizing the quadratic error between the numerical and experimental results of the vibration response caused by system unbalance. There are several investigations on methods for identifying unbalance and bearing dynamic parameters [11–15]. Recently, Wang et al. [16] presented the development of algorithms for the simultaneous identification of unbalance and bearing dynamic parameters. In both cases, the proposed algorithms were validated by comparison with experimental data. Additionally, in [17], the authors estimated the rotordynamic coefficients of a controllable floating ring bearing with a magnetorheological fluid (MRF) showing that the magnetic field-induced, field-dependent viscosity of the MRF changes the stiffness and damping bearing coefficients, which can be used to modify the dynamic behavior of the rotor-bearing system. In 2020, Kang et al. [18] used the Kalman filter to estimate the bearing dynamic coefficients of a flexible rotor-bearing system. The rotor system is modeled with Timoshenko beam elements, but the imbalance force considered in the dynamic model is calculated for a constant rotational velocity condition. More recently, in 2021, Chen et al. [19] proposed a method to simultaneously identify the parameters of the oil-film bearings and active magnetic bearings/bearingless motors AMBs/BELMs along with the residual unbalanced forces during the unbalanced vibration of the rotor. The proposed method requires independent rotor responses and control currents to form a regression equation to estimate all of the unknown parameters. Independent rotor responses are realized by changing the PID control parameters of the AMBs/BELMs. The finite element method is used to model the system by using Timoshenko beam elements, and both numerical and experimental results are presented at a unique operation velocity of 2400 rpm. Taherkhani and Ahmadian [20] used the Bayesian approach to an appropriate parameter selection procedure and suitable sampling strategy for stochastic model updating to investigate variability in the dynamic behavior of a complex turbo compressor supported by hydrodynamic bearings, leading to successful parameter identification results. Brito Jr. et al. [21] presented an experimental method to estimate the direct and cross-coupled dynamic coefficients of tilting-pad journal bearings of vertical hydro-generators. The method employs only the shaft radial relative vibrations, and the bearing radial absolute vibrations originated by the hydro-generator residual unbalance. The authors affirmed that the vibration measurements required by the estimation method could be a major problem in low-speed machines (less than 400 rpm). Although any type of bearing provides stiffness and damping forces that may depend on the operation speed and many other factors, linearized force coefficients are widely used to model the reaction forces from fluid film bearings. These linear force coefficients are derived from the assumption of small amplitude motions about an equilibrium position [22] and have been used to study the dynamic responses and analyze the stability of rotor systems supported by oil-lubricated tilting-pad bearings, cylindrical bearings and foil bearings, as pointed out in [23] and references within. Recently, Dyk et al. [24] presented diverse linearization methods in the stability analysis of rotating systems supported on floating ring bearing (FRBs), demonstrating the usefulness of the linear force coefficients to predict the dynamic behavior of non-linear systems such as turbochargers supported by FRBs.

There is also substantial literature on parameter identification and estimation methods. Most of these schemes are essentially asymptotic, recursive or complex [25–27], and, according to Arias-Montiel et al. [28], these methods lead to unrealistic implementations. Over the past few years, another method of parametric identification called algebraic identification has been successfully implemented in a wide range of engineering applications [29]. The algebraic identification method is based on differential algebra and operational calculus for developing estimators in determining unknown system parameters from a mathematical model. These estimations are carried out on-line in continuous or discrete time. An advantage of algebraic identification over other methods is that it provides identification expressions that are completely independent of the initial system conditions. Algebraic identification has been used for parameter and signal estimation in linear and non-linear vibrational mechanical systems [30–39]. Numerical and experimental results show that algebraic identification is extremely robust against parameter uncertainty, frequency variations, measurement errors and signal noise. Additional information on the algebraic identification robustness and other advantages and disadvantages of this method are highlighted by Sira-Ramírez et al. in [29].

In this work, a novel methodology for developing two mathematical models for identifying the unknown stiffness and damping parameters of bearings in multiple degree-of-freedom (DOF) rotor-bearing systems is proposed. This methodology is based on the algebraic identification technique. Developed identifiers are obtained based on an FE model for a multiple DOF rotor-bearing system that considers the effects of rotational inertia, gyroscopic effects, shear deformations, internal damping and linear forces attributable to stiffness and damping parameters of the supports. Estimators are developed for two different operation conditions of the rotor-bearing system: constant and variable rotational speed. Analysis and evaluation of the proposed identifiers is carried out by numerical results showing the viability for applying algebraic identification techniques for the rotordynamic coefficients in rotor-bearing systems.

2. Materials and Methods

2.1. Mathematical Model of the Rotor-Bearing System

The FE method is used to obtain the mathematical model of the multiple DOF rotor-bearing system. The shaft is modelled with a finite element type beam with four DOF per node, two lateral displacements and two rotations (beam deflections), as illustrated in Figure 1.

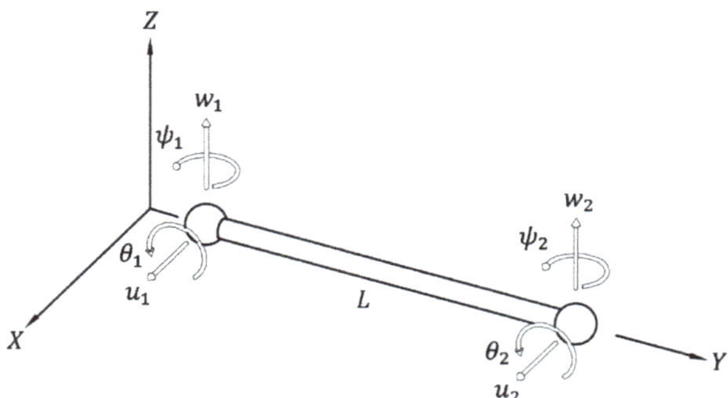

Figure 1. Beam-like finite element for the modelling of the rotor-bearing shaft.

The nodal displacement vector is defined as

$$\{\delta\} = \{u_1, w_1, \psi_1, \theta_1, u_2, w_2, \psi_2, \theta_2\}^T \tag{1}$$

where superscript T denotes the transposed vector.

Displacement and rotations corresponding to the movement along X and Z directions are

$$\begin{aligned} \{\delta_u\} &= \{u_1, \psi_1, u_2, \psi_2\}^T \\ \{\delta_w\} &= \{w_1, \theta_1, w_2, \theta_2\}^T \end{aligned} \tag{2}$$

The mathematical model of the multiple DOF rotor-bearing system with excitation by unbalanced mass is given by [2]

$$[M]\{\ddot{\delta}\} + \left[C\left(\dot{\phi}\right)\right]\{\dot{\delta}\} + \left[K\left(\ddot{\phi}\right)\right]\{\delta\} = \dot{\phi}^2\left\{F_{u(1)}(\phi)\right\} + \ddot{\phi}\left\{F_{u(2)}(\phi)\right\} \tag{3}$$

with

$$\begin{aligned} F_{u(1)} &= m_u d(\sin(\phi + \alpha) + \cos(\phi + \alpha)) \\ F_{u(2)} &= m_u d(\sin(\phi + \alpha) - \cos(\phi + \alpha)) \end{aligned}$$

where m_u, d and α, are mass, eccentricity and angular position of system unbalance, respectively, $\ddot{\phi}$ and $\dot{\phi}$ are angular acceleration and velocity of the rotor-bearing system, respectively, and $\phi = \dot{\phi}t$. Moreover, $\{\delta\}$ is a vector with all the nodal displacements, $[M]$ is the global mass matrix of the system, $\left[C\left(\dot{\phi}\right)\right]$ is the global damping matrix that includes gyroscopic effects as a function of the rotational velocity $\left(\dot{\phi}[C_2]\right)$ and $[C_1]$ that represents the damping in the supports, $\left[K\left(\ddot{\phi}\right)\right]$ is the global stiffness matrix constituted by $[K_1]$, $[K_2]$, which include the supports and rotor stiffness, respectively, and $\ddot{\phi}[K_3]$, which is a stiffness term as a function of the rotational acceleration of the system. Finally, $\left\{F_{u(1)}(\phi)\right\}$ and $\left\{F_{u(2)}(\phi)\right\}$ are the components of the centrifugal force vector caused by the unbalanced mass. Shape functions for the beam type finite element and a detailed definition for matrices in Equation (3) are provided in Appendix A.

Stiffness and damping matrices provided by the bearings are obtained by determining the generalized forces that these elements exert on the rotor shaft. After applying the virtual work principle to the bearing model shown in Figure 2, forces acting on the rotor can be expressed in a matrix form as [40]

$$\begin{Bmatrix} F_{u_i} \\ F_{w_i} \end{Bmatrix} = -\begin{bmatrix} k_{xx} & k_{xz} \\ k_{zx} & k_{zz} \end{bmatrix}\begin{Bmatrix} u_i \\ w_i \end{Bmatrix} - \begin{bmatrix} c_{xx} & c_{xz} \\ c_{zx} & c_{zz} \end{bmatrix}\begin{Bmatrix} \dot{u}_i \\ \dot{w}_i \end{Bmatrix} \tag{4}$$

where *i* denotes the nodal location of the bearing inside the rotordynamic systems. Matrices from the right side of Equation (4) are stiffness and damping matrices corresponding to system supports $[K_1]$ and $[C_1]$, respectively.

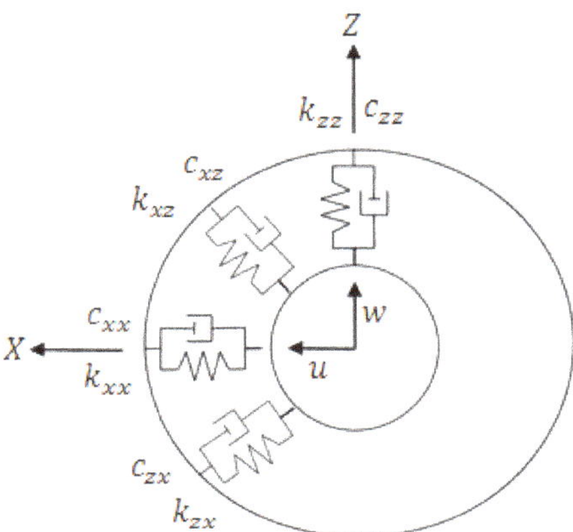

Figure 2. Stiffness and damping parameters in bearings [40].

2.2. Operation Velocity of the Rotor-Bearing System

Two different conditions for the operation velocity of the rotor-bearing system are considered: constant velocity and a linear ramp excitation.

Under the constant velocity scheme, no time variation of the rotating machine excitation is considered. This condition can be expressed as

$$\dot{\phi}(t) = \Omega = \text{constant} \tag{5}$$

The term "ramp excitation" means a continuous variation in the excitation frequency with a specific ratio with respect to time and can be ascendant (up) or descendent (down). With most real rotating systems, the excitation frequency does not change in a linear manner with respect to time. However, in some cases, frequency variation is sufficiently slow to be approximated by a linear function. For the solution of Equation (3), it is considered a variation of the excitation frequency of the form

$$\dot{\phi}(t) = \dot{\phi}_0 + \ddot{\phi}t \tag{6}$$

where:

$\dot{\phi}_0$ is the excitation frequency at the ramp beginning;

$\ddot{\phi}$ is the change ratio with respect to time of the excitation frequency;

t is the time.

2.3. Mathematical Model for Bearing Rotordynamic Parameters Identification

The development of the mathematical model of the identifier is carried out from the rotordynamic system model given in Equation (3), considering both cases: constant and variable system operation velocity.

2.3.1. Algebraic Identifier with Constant Operation Velocity

As pointed out above, it is necessary to have a mathematical model for the dynamic behavior of the rotor-bearing system to develop algebraic identifiers. From this model and through an algebraic manipulation of the equations, estimators for the unknown parameters are obtained.

If a constant rotational velocity of the system is considered, Equation (3) can be written as

$$[M]\{\ddot{\delta}\} + [C_1 + \Omega C_2]\{\dot{\delta}\} + [K_1 + K_2]\{\delta\} = m_u d\Omega^2 \{\sin(\Omega t + \alpha) + \cos(\Omega t + \alpha)\} \tag{7}$$

Now, Equation (7) is multiplied by t^2 and, after that, the result is twice integrated with respect to time, giving

$$\int^{(2)} [K_1]t^2\{\delta\} + [C_1]\left[\int t^2\{\delta\} - 2\int^{(2)} t\{\delta\}\right] = \int^{(2)} m_u d\Omega^2\{\sin(\Omega t + \alpha) + \cos(\Omega t + \alpha)\}t^2 \tag{8}$$

where $\int^{(2)} f(t)$ denotes iterated integrals. Furthermore, bearing stiffness and damping terms to be identified are included in $[K_1]$ and $[C_1]$, respectively. Therefore, after the integration of the left side of Equation (8) and an algebraic treatment, the following expression can be obtained

$$\int^{(2)} \left[[M]\{\ddot{\delta}\} + [C_1 + \Omega C_2]\{\dot{\delta}\} + [K_1 + K_2]\{\delta\}\right]t^2$$
$$= \int^{(2)} [2\Omega C_2 t - 2M - K_2 t^2]\{\delta\} + \int [4M - \Omega C_2 t]t\{\delta\} \tag{9}$$
$$-\{M\}t^2\{\delta\} + \int^{(2)} m_u d\Omega^2\{\sin(\Omega t + \alpha) + \cos(\Omega t + \alpha)\}t^2$$

Equation (9) can be separated into individual equation systems for each node where the bearings are located. These equations can be presented in the form

$$\begin{bmatrix} k_{xx} & k_{xz} \\ k_{zx} & k_{zz} \end{bmatrix} \int^{(2)} t^2 \begin{Bmatrix} u_i \\ w_i \end{Bmatrix} + \begin{bmatrix} c_{xx} & c_{xz} \\ c_{zx} & c_{zz} \end{bmatrix} \left(\int t^2 \begin{Bmatrix} u_i \\ w_i \end{Bmatrix} - 2\int^{(2)} t \begin{Bmatrix} u_i \\ w_i \end{Bmatrix}\right) = \begin{Bmatrix} b_{ui} \\ b_{wi} \end{Bmatrix} \tag{10}$$

To solve Equation (10) an equal number of equations and unknowns is needed. For this, Equation (10) is successively integrated three times in order to obtain the missing equations, which are written as

$$\begin{bmatrix} k_{xx} & k_{xz} \\ k_{zx} & k_{zz} \end{bmatrix} \int^{(3)} t^2 \begin{Bmatrix} u_i \\ w_i \end{Bmatrix} + \begin{bmatrix} c_{xx} & c_{xz} \\ c_{zx} & c_{zz} \end{bmatrix} \left(\int^{(2)} t^2 \begin{Bmatrix} u_i \\ w_i \end{Bmatrix} - 2\int^{(3)} t \begin{Bmatrix} u_i \\ w_i \end{Bmatrix}\right) = \int \begin{Bmatrix} b_{ui} \\ b_{wi} \end{Bmatrix} \tag{11}$$

$$\begin{bmatrix} k_{xx} & k_{xz} \\ k_{zx} & k_{zz} \end{bmatrix} \int^{(4)} t^2 \begin{Bmatrix} u_i \\ w_i \end{Bmatrix} + \begin{bmatrix} c_{xx} & c_{xz} \\ c_{zx} & c_{zz} \end{bmatrix} \left(\int^{(3)} t^2 \begin{Bmatrix} u_i \\ w_i \end{Bmatrix} - 2\int^{(4)} t \begin{Bmatrix} u_i \\ w_i \end{Bmatrix}\right) = \int^{(2)} \begin{Bmatrix} b_{ui} \\ b_{wi} \end{Bmatrix} \tag{12}$$

$$\begin{bmatrix} k_{xx} & k_{xz} \\ k_{zx} & k_{zz} \end{bmatrix} \int^{(5)} t^2 \begin{Bmatrix} u_i \\ w_i \end{Bmatrix} + \begin{bmatrix} c_{xx} & c_{xz} \\ c_{zx} & c_{zz} \end{bmatrix} \left(\int^{(4)} t^2 \begin{Bmatrix} u_i \\ w_i \end{Bmatrix} - 2\int^{(5)} t \begin{Bmatrix} u_i \\ w_i \end{Bmatrix}\right) = \int^{(3)} \begin{Bmatrix} b_{ui} \\ b_{wi} \end{Bmatrix} \tag{13}$$

From Equations (10)–(13), a linear system equation is obtained for each node where the bearings are located. These equations can be expressed as

$$[A_s(t)]\{\Theta_s\} = \{b_s(t)\} \tag{14}$$

where $\{\Theta_s\} = \{k_{xx}\ k_{xz}\ k_{zx}\ k_{zz}\ c_{xx}\ c_{xz}\ c_{zx}\ c_{zz}\}^T$ denotes the transposed vector of parameters to be identified and $[A_s(t)]$, $\{b_s(t)\}$ are 8×8 and 8×1, respectively.

As can be observed in Figure 2, eight parameters are required to define stiffness and damping characteristics provided by the system supports. This is because in order to obtain the terms of $[A_s(t)]$ and $\{b_s(t)\}$ in Equation (14), eight simultaneous equations involving the unknown support parameters are required to obtain their magnitudes.

From Equation (14) it can be concluded that vector $\{\Theta_s\}$ is identifiable if, and only if, the dynamic system trajectory is persistent. That is to say, the trajectories or dynamic system behaviors satisfy the condition $det[A_s(t)] \neq 0$. In general, this condition is maintained at least in a small interval $(t_0, t_0 + \epsilon]$ where ϵ is a positive and sufficiently small value [29]. Then, the linear system Equation (14) is solved to obtain the algebraic identifier for determining the stiffness and damping parameters of rotor-bearing support with constant operation velocity.

$$\{\Theta_s\} = [A_s]^{-1}\{b_s\} \ \forall t \in (t_0, t_0 + \epsilon]. \tag{15}$$

It is important to mention that to identify the support parameters, lateral vibration measurements at the node and the nodal slopes are required. Moreover, similar information from the adjacent node is also needed. The nodal slopes can be calculated by numerical approximation using the lateral displacements from two adjacent nodes.

2.3.2. Algebraic Identifier with Variable Operation Velocity

In this section, the rotor-bearing system velocity is considered as a linear ramp of excitation. The mathematical model of the system is defined by Equation (3). In order to develop the parameter identifier, this equation is rewritten as follows

$$[M]\{\ddot{\delta}\} + \left[C_1 + \dot{\phi}C_2\right]\{\dot{\delta}\} + \left[K_1 + K_2 + \ddot{\phi}K_3\right]\{\delta\} = \dot{\phi}^2 F_1(\phi) + \ddot{\phi}F_2(\phi) \tag{16}$$

By multiplying Equation (16) by t^2 and integrating the result twice with respect to time, the following is obtained

$$\int^{(2)}\left[[M]\{\ddot{\delta}\} + \left[C_1 + \dot{\phi}C_2\right]\{\dot{\delta}\} + \left[K_1 + K_2 + \ddot{\phi}K_3\right]\{\delta\}\right]t^2 = \int^{(2)}\left\{\dot{\phi}^2 F_1(\phi) + \ddot{\phi}F_2(\phi)\right\}t^2 \tag{17}$$

where $\int^{(2)} \varphi(t)$ are iterated time-integrals of the form $\int_0^t \int_0^{\sigma_1} \cdots \int_0^{\sigma_{n-1}} \varphi(\sigma_n)d\sigma_n \cdots d\sigma_1$ with $\int \varphi(t) = \int_0^t \varphi(\sigma)d\sigma$, and n a positive integer.

Similarly for the case of constant velocity, matrices $[K_1]$ and $[C_1]$ contain the stiffness and damping parameters provided by the supports. Therefore, after the integration of the left part of Equation (17) and rearranging terms, we have

$$\begin{aligned}
\int^{(2)}[K]_1 t^2\{\delta\} + [C_1]\left[\int t^2\{\delta\} - 2\int^{(2)} t\{\delta\}\right] &= \int\left[4Mt - \dot{\phi}C_2 t^2\right]\{\delta\} \\
&+ \int^{(2)}\left[C_2\left(\ddot{\phi}t^2 + 2\dot{\phi}t\right) - 2M - \left(K_2 + \ddot{\phi}K_3\right)t^2\right]\{\delta\} - [M]t^2\{\delta\} \\
&+ \int^{(2)}\left\{\dot{\phi}^2 F_1(\phi) + \ddot{\phi}F_2(\phi)\right\}t^2
\end{aligned} \tag{18}$$

It is worth mentioning that Equation (18) can be separated into individual equation systems for each node where the bearings are located. These equations can be written as follows

$$\begin{bmatrix} k_{xx} & k_{xz} \\ k_{zx} & k_{zz} \end{bmatrix}\int^{(2)} t^2\begin{Bmatrix} u_i \\ w_i \end{Bmatrix} + \begin{bmatrix} c_{xx} & c_{xz} \\ c_{zx} & c_{zz} \end{bmatrix}\left(\int t^2\begin{Bmatrix} u_i \\ w_i \end{Bmatrix} - 2\int^{(2)} t\begin{Bmatrix} u_i \\ w_i \end{Bmatrix}\right) = \begin{Bmatrix} b_{ui} \\ b_{wi} \end{Bmatrix} \tag{19}$$

To solve Equation (19), an equal number of equations and unknows is required. Therefore, Equation (19) is successively integrated three times to obtain the missing equations which are expressed as

$$\begin{bmatrix} k_{xx} & k_{xz} \\ k_{zx} & k_{zz} \end{bmatrix}\int^{(3)} t^2\begin{Bmatrix} u_i \\ w_i \end{Bmatrix} + \begin{bmatrix} c_{xx} & c_{xz} \\ c_{zx} & c_{zz} \end{bmatrix}\left(\int^{(2)} t^2\begin{Bmatrix} u_i \\ w_i \end{Bmatrix} - 2\int^{(3)} t\begin{Bmatrix} u_i \\ w_i \end{Bmatrix}\right) = \int\begin{Bmatrix} b_{ui} \\ b_{wi} \end{Bmatrix} \tag{20}$$

$$\begin{bmatrix} k_{xx} & k_{xz} \\ k_{zx} & k_{zz} \end{bmatrix}\int^{(4)} t^2\begin{Bmatrix} u_i \\ w_i \end{Bmatrix} + \begin{bmatrix} c_{xx} & c_{xz} \\ c_{zx} & c_{zz} \end{bmatrix}\left(\int^{(3)} t^2\begin{Bmatrix} u_i \\ w_i \end{Bmatrix} - 2\int^{(4)} t\begin{Bmatrix} u_i \\ w_i \end{Bmatrix}\right) = \int^{(2)}\begin{Bmatrix} b_{ui} \\ b_{wi} \end{Bmatrix} \tag{21}$$

$$\begin{bmatrix} k_{xx} & k_{xz} \\ k_{zx} & k_{zz} \end{bmatrix}\int^{(5)} t^2\begin{Bmatrix} u_i \\ w_i \end{Bmatrix} + \begin{bmatrix} c_{xx} & c_{xz} \\ c_{zx} & c_{zz} \end{bmatrix}\left(\int^{(4)} t^2\begin{Bmatrix} u_i \\ w_i \end{Bmatrix} - 2\int^{(5)} t\begin{Bmatrix} u_i \\ w_i \end{Bmatrix}\right) = \int^{(3)}\begin{Bmatrix} b_{ui} \\ b_{wi} \end{Bmatrix} \tag{22}$$

From Equations (19)–(22), a linear system equation is obtained for each node where the bearings are located. These equations can be expressed as

$$[A_s(t)]\{\Theta_s\} = \{b_s(t)\} \tag{23}$$

where $\{\Theta_s\} = \{k_{xx}\, k_{xz}\, k_{zx}\, k_{zz}\, c_{xx}\, c_{xz}\, c_{zx}\, c_{zz}\}^T$ denotes the transposed vector of parameters to be identified and $[A_s(t)]$, $\{b_s(t)\}$ are 8×8 and 8×1, respectively.

Again, the condition $det[A_s(t)] \neq 0$ must be satisfied to identify the vector $\{\Theta_s\}$.

From the solution of Equation (23), a mathematical model for an on-line identifier of stiffness and damping bearing parameters can be obtained as

$$\{\Theta_s\} = \left[A_s^{-1} \right] \{b_s\} \quad \forall t \in (t_0, t_0 + \epsilon]$$

(24)

As can be observed, algebraic identification of stiffness and damping bearing parameters is independent of system initial conditions and only depends on the displacement vector and the type of ramp excitation. It is important to mention that as with the case of constant velocity, to identify the supports parameters, lateral vibration measurements at the node and the nodal slopes are required. Moreover, similar information from the adjacent node is also needed.

3. Results

In Figure 3, a scheme of the rotor-bearing system considered in this work and its discretization is presented.

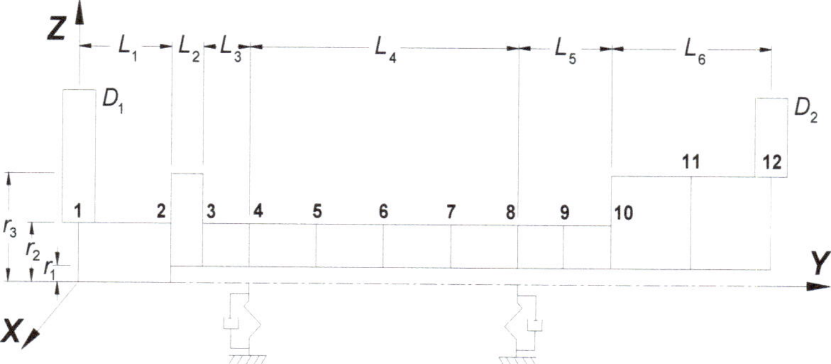

Figure 3. Rotor-bearing system scheme [40].

To obtain the mathematical model for the rotor-bearing system using the FE method, it was discretized into 11 beam-like elements, as is shown in Figure 3. The system includes two inertial disks located at nodes 1 and 12, while supports (bearings) are placed at nodes 4 and 8. The correct nodal location ensures that the simulation replicates the model's real conditions. In addition, two unbalanced masses were considered in two different angular positions located on inertial disks D_1 and D_2.

In Table 1, the mechanical and geometrical properties of the shaft are shown, while the inertial properties of discs and unbalanced masses are presented in Table 2.

Table 1. Mechanical and geometrical properties of the rotor-bearing shaft.

Parameter	Value	Parameter	Value
Modulus of elasticity	2×10^{11} N/m^2	L_1	0.035 m
Density	7800 kg/m^3	L_2	0.010 m
Poisson ratio	0.3	L_3	0.025 m
r_1	0.005 m	L_4	0.130 m
r_2	0.02 m	L_5	0.050 m
r_3	0.035 m	L_6	0.050 m

Table 2. Inertial properties of the disks and unbalance masses.

Parameter	Value	Parameter	Value
D_1 mass	1.2 kg	D_2 mass	1.0 kg
D_1 moment of inertia	1.2×10^{-3} kg·m^2	D_2 moment of inertia	1.0×10^{-3} kg·m^2
D_1 polar moment of inertia	2.4×10^{-3} kg·m^2	D_2 polar moment of inertia	2.0×10^{-3} kg·m^2
D_1 mass unbalance	5×10^{-7} kg·m $\angle 0\ rad$	D_2 mass unbalance	5×10^{-7} kg·m $\angle \pi\ rad$

In Table 3, the stiffness and damping bearing parameters [40] used for numerical simulation are presented.

Table 3. Stiffness and damping bearing parameters [40].

Parameter	Bearing 1 (Node 4)	Bearing 2 (Node 8)
k_{xx}	8×10^7 N/m	5×10^7 N/m
k_{xz}	-1×10^7 N/m	-2×10^7 N/m
k_{zx}	-6×10^7 N/m	-4×10^7 N/m
k_{zz}	1×10^8 N/m	7×10^7 N/m
c_{xx}	8×10^3 N·s/m	6×10^3 N·s/m
c_{xz}	-3×10^3 N·s/m	-1.5×10^3 N·s/m
c_{zx}	-3×10^3 N·s/m	-1.5×10^3 N·s/m
c_{zz}	1.2×10^4 N·s/m	8×10^3 N·s/m

On-line algebraic identification of stiffness and damping bearing parameters was determined based on the vibratory response of the rotor-bearing system in the time domain, which was obtained from Equations (3) and (7) by using the Newmark method for numerical integration.

3.1. Algebraic Parameter Identification with Constant System Velocity

The displacement vector used in the algebraic identification procedure was obtained from Equation (7) by using the Newmark method for numerical integration and taking into account a constant rotational velocity of the rotor-bearing system.

In Figure 4, vibration signals at node 4 (corresponding to bearing 1 location) of the rotor-bearing system of Figure 3 are presented. This response is obtained for an operation rotational velocity $\Omega = 600$ rpm. These signals, the nodal slopes and the corresponding information of the nodes adjacent to the bearing locations are the required data to identify stiffness and damping parameters of the bearings.

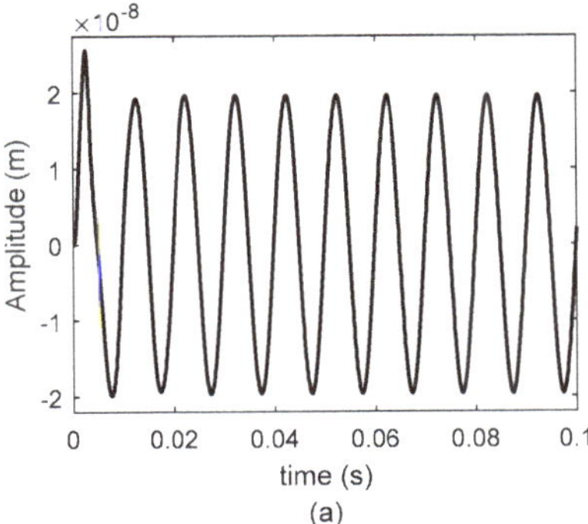
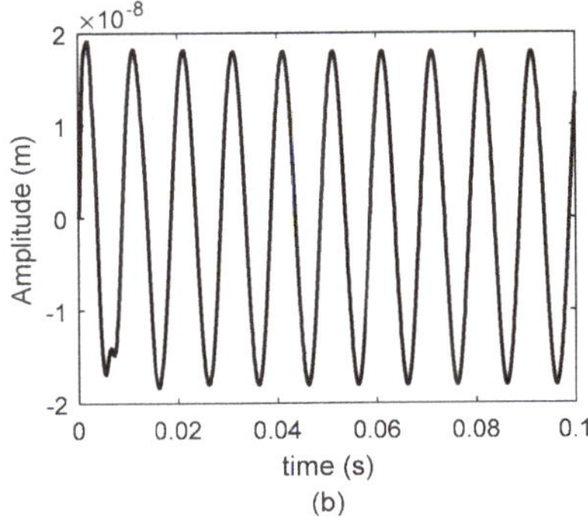

Figure 4. Vibration signal at node 4 (bearing 1) at 600 rpm: (**a**) X direction; (**b**) Z direction.

Figures 5 and 6 present the obtained results from the numerical simulation for the algebraic identification of stiffness and damping parameters for bearing 1, while Figures 7 and 8 show the results corresponding to bearing 2. It is important to mention that the sample time used in the simulation was 0.1 milliseconds. However, by carrying out numerical simulations with different sample times, it was observed that the shorter the sampling period, the faster the identifier converges.

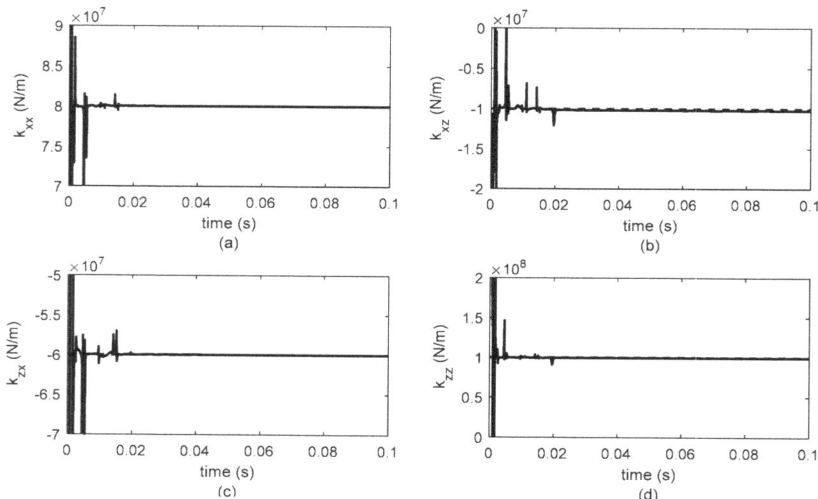

Figure 5. Identified stiffness parameters for bearing 1 at 600 rpm. (**a**) k_{xx}, (**b**) k_{xz}, (**c**) k_{zx}, (**d**) k_{zz}.

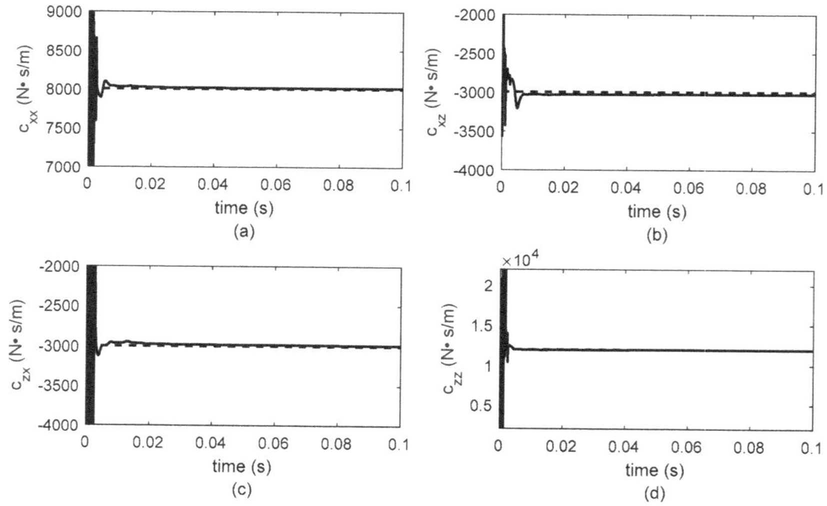

Figure 6. Identified damping parameters for bearing 1 at 600 rpm. (**a**) c_{xx}, (**b**) c_{xz}, (**c**) c_{zx}, (**d**) c_{zz}.

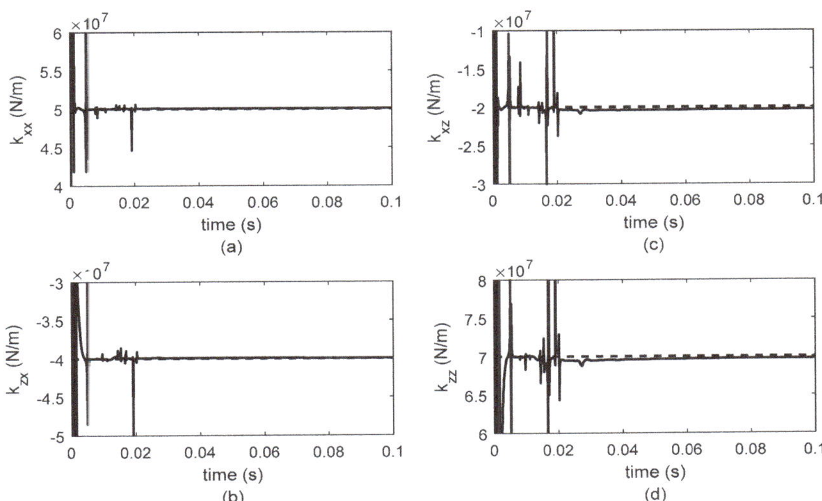

Figure 7. Identified stiffness parameters for bearing 2 at 600 rpm. (**a**) k_{xx}, (**b**) k_{xz}, (**c**) k_{zx}, (**d**) k_{zz}.

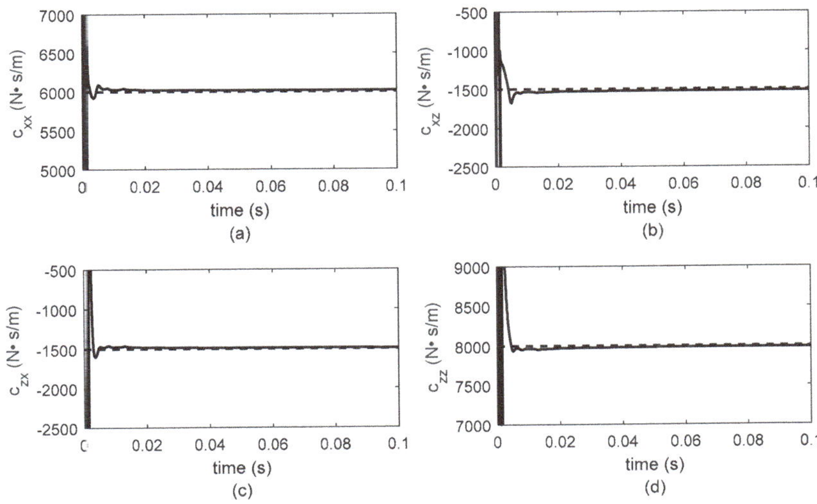

Figure 8. Identified damping parameters for bearing 2 at 600 rpm. (**a**) c_{xx}, (**b**) c_{xz}, (**c**) c_{zx}, (**d**) c_{zz}.

As can be observed in Figures 5–8, the identification of both stiffness and damping parameters of the bearings is carried out in less than 0.1 s, and once the parameter reaches the identified value, this remains for the rest of the time period. For a better analysis of the identifier behavior, only results for 0.1 s are presented in Figures 5–8, because it is important to observe the time that the identifier requires to converge to the estimated value.

3.2. Algebraic Parameter Identification with Variable System Velocity

The displacement vector used as input data for the algebraic identification is obtained from Equation (3) by using the Newmark method for numerical integration and taking into account a linear ramp excitation with angular acceleration $\ddot{\phi} = 10\ rad/s^2$. The rotor-bearing system response at node 4 is shown in Figure 9 where the vibratory behavior of the system in the location of bearing 1 can be appreciated.

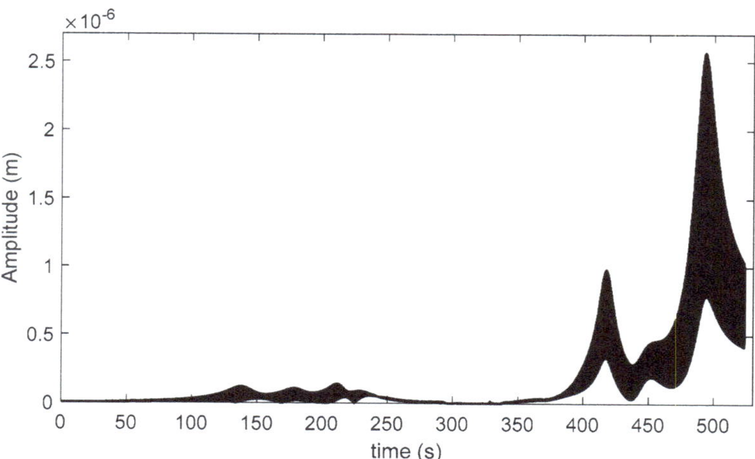

Figure 9. System vibratory response at node 4 with a linear ramp of excitation of $10 \ rad/s^2$.

In Figures 10–13, the behavior of the algebraic identifier for bearing stiffness and damping parameters of both bearings (placed at nodes 4 and 8) is shown as a function of time.

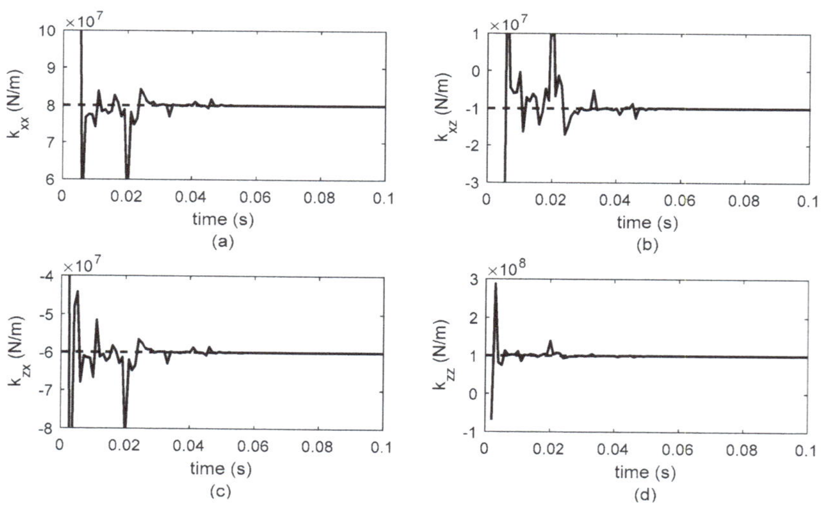

Figure 10. Identified stiffness parameters for bearing 1 with a linear ramp of excitation of $10 \ rad/s^2$. (a) k_{xx}, (b) k_{xz}, (c) k_{zx}, (d) k_{zz}.

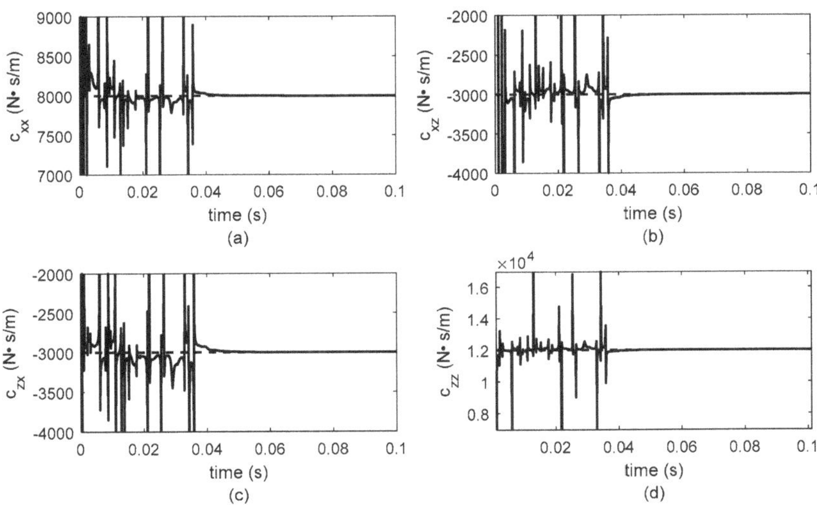

Figure 11. Identified damping parameters for bearing 1 with a linear ramp of excitation of $10\ rad/s^2$. (**a**) c_{xx}, (**b**) c_{xz}, (**c**) c_{zx}, (**d**) c_{zz}.

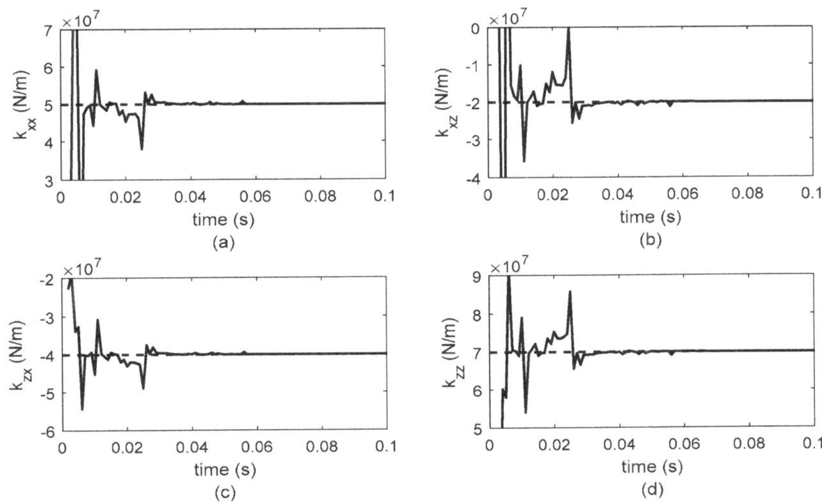

Figure 12. Identified stiffness parameters for bearing 2 with a linear ramp of excitation of $10\ rad/s^2$. (**a**) k_{xx}, (**b**) k_{xz}, (**c**) k_{zx}, (**d**) k_{zz}.

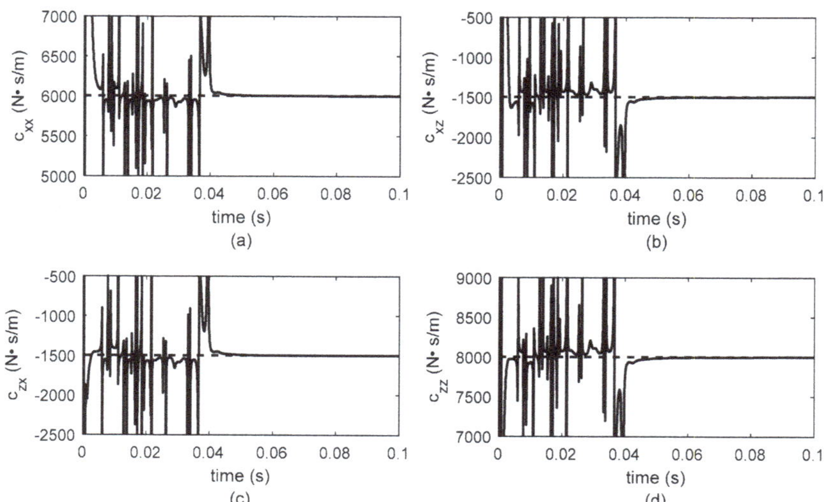

Figure 13. Identified damping parameters for bearing 2 with a linear ramp of excitation of 10 rad/s^2. (**a**) c_{xx}, (**b**) c_{xz}, (**c**) c_{zx}, (**d**) c_{zz}.

As in the case of constant rotor system velocity, stiffness and damping bearing parameters are identified in les tan 0.1 s as can be observed in Figures 10–13. Furthermore, the parameter values remain constant until the rotor-bearing system reaches its nominal operation velocity. For a better analysis of the identifier behavior, the results for 0.1 s are presented in Figures 10–13 because it is important to observe the required time for the identifier convergence. The sample time used to solve Equation (3) using the Newmark method was 0.1 milliseconds. The numerical solution of Equation (3) was used as input data for the proposed algebraic identifier. Moreover, achieving this sample time with diverse data acquisition systems for experimental implementation was verified.

4. Discussion

Different numerical simulations were carried out in order to determine the robustness of the proposed identifiers under different conditions for the rotor-bearing system velocity. For the constant velocity case, different magnitudes for the rotor system velocity were considered, while for the variable velocity case, different ramps of excitation were explored.

Figure 14 shows results for the algebraic identification of stiffness and damping parameters for bearing 1 at a constant operation velocity of the rotor-bearing system of 50,000 rpm. A rapid identifier convergence to the estimated values can be observed, meaning that an increase in operation velocity does not affect the identifier performance. It is important to mention that, while the results for the identification of damping parameters of bearing 1 and the stiffness and damping parameters of bearing 2 are not presented, these parameters are correctly identified in less than 0.1 s.

The identifier performance for different ramps' excitation was analyzed. The acceleration values considered for numerical simulation were: $\ddot{\phi} = 10 \; rad/s^2$, $\ddot{\phi} = 100 \; rad/s^2$, $\ddot{\phi} = 1000 \; rad/s^2$, $\ddot{\phi} = 3000 \; rad/s^2$ and $\ddot{\phi} = 6000 \; rad/s^2$. The result for $\ddot{\phi} = 10 \; rad/s^2$ were reported in the previous section. Due to the similar behavior of the identifier with different acceleration values, only results for $\ddot{\phi} = 6000 \; rad/s^2$ are shown here. The rotor-bearing system response for a ramp of excitation with the mentioned value of acceleration at node 4 (bearing 1 location) is presented in Figure 15. It can be seen that there is a considerable change in the time scale in comparison with Figure 9 because the acceleration is increased 600 times.

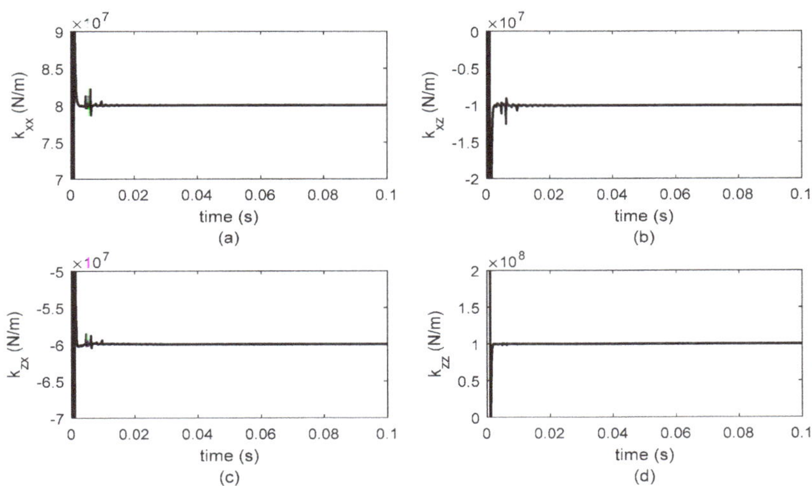

Figure 14. Identified stiffness parameters for bearing 1. (**a**) k_{xx}, (**b**) k_{xz}, (**c**) k_{zx}, (**d**) k_{zz} at 50,000 rpm.

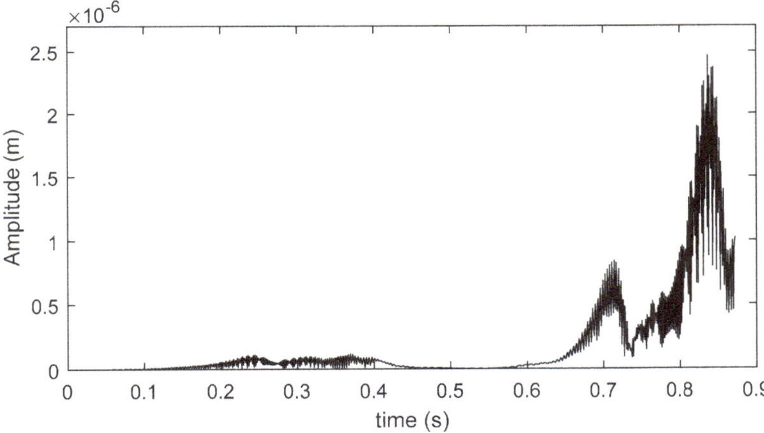

Figure 15. System vibratory response at node 4 with a linear ramp excitation with acceleration of $6000\ rad/s^2$.

The algebraic identification performance under the conditions described above is shown in Figures 16–19 where the estimation for the stiffness and damping bearings parameter is visualized. For this simulation, the system response in Figure 15 is used as input data.

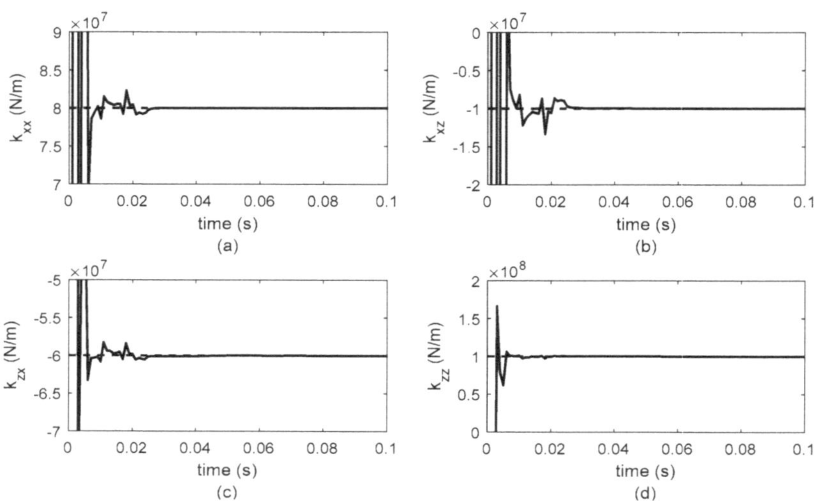

Figure 16. Identified stiffness parameters for bearing 1 with a ramp excitation of 6000 rad/s^2. (**a**) k_{xx}, (**b**) k_{xz}, (**c**) k_{zx}, (**d**) k_{zz}.

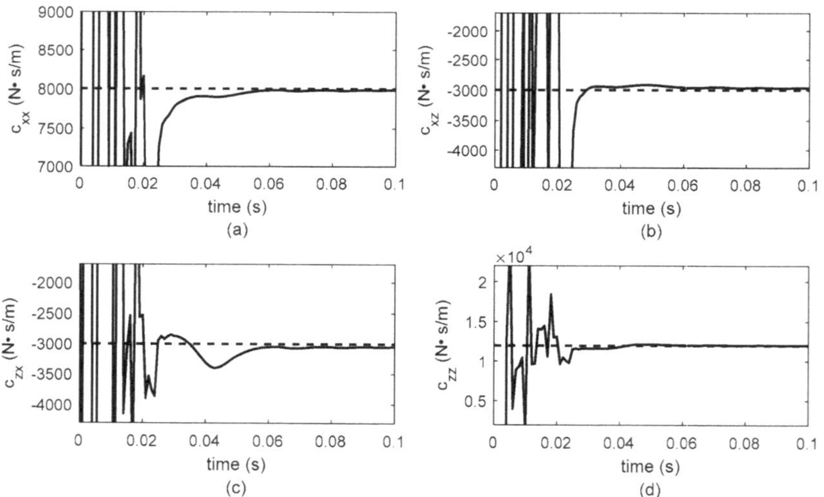

Figure 17. Identified damping parameters for bearing 1 with a ramp excitation of 6000 rad/s^2. (**a**) c_{xx}, (**b**) c_{xz}, (**c**) c_{zx}, (**d**) c_{zz}.

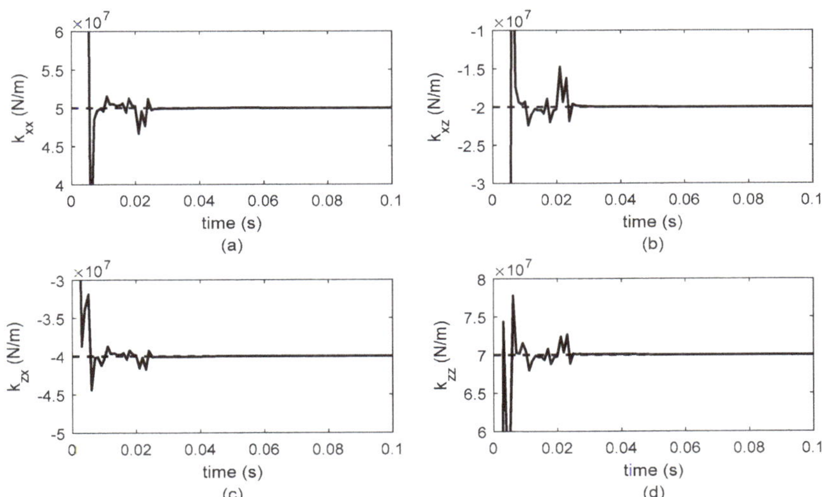

Figure 18. Identified stiffness parameters for bearing 2 with a ramp excitation of 6000 rad/s^2. (**a**) k_{xx}, (**b**) k_{xz}, (**c**) k_{zx}, (**d**) k_{zz}.

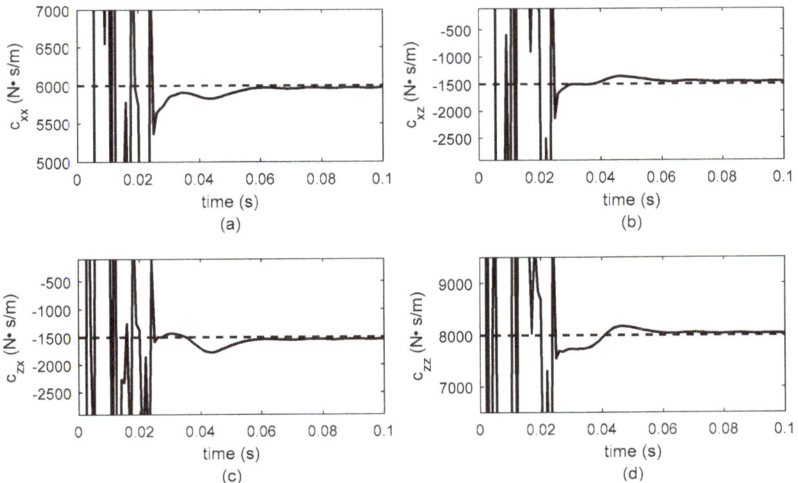

Figure 19. Identified damping parameters for bearing 2 with a ramp excitation of 6000 rad/s^2. (**a**) c_{xx}, (**b**) c_{xz}, (**c**) c_{zx}, (**d**) c_{zz}.

From the results presented in Figures 16–19, it can be observed that despite the linear ramp excitation being 600 times faster than the corresponding one in Figure 9, the proposed identifier rapidly converges to the estimated values and remains in these values for the rest of the time period. Note that the algebraic identifier is not affected by the system's acceleration and only depends on the displacement vector at each instance. The robustness of the algebraic identification method against acceleration ramp variations had already been proved by Mendoza-Larios et al. [36] but only for the identification of unbalance parameters in rotor-bearing parameters.

Furthermore, the obtained results for both cases, constant and variable rotor-bearing velocity, show a transient state of the identifiers before the convergence to the estimated values of stiffness and damping bearing parameters. This behavior is due to the sample time used in the numerical simulations of the identifiers for solving the iterated integrals of Equations (15) and (24), which utilize the trapezoidal rule. According to Kharab and Guenther [41], this method presents major calculation errors in comparison with other integration methods. However, it was found that the smaller the sample time the shorter the error in the trapezoidal rule calculation.

5. Conclusions

In this article the identification problem for stiffness and damping parameters of the supports in rotor-bearing systems was addressed. The system model was obtained by the finite element method and using a finite element type beam, which consider the effects of rotational inertia, gyroscopic moments and shear deformations. The algebraic identification technique was applied to the finite element model to obtain two identifiers for the stiffness and damping parameters attributable to the bearings. The first identifier considers a rotor-bearing system operating at a constant velocity, and the second with a linear ramp of excitation as a system velocity input. The numerical results present the identifier behavior showing a fast convergence and robustness in both operation conditions with different values of constant rotational velocity and ramp of acceleration. The numerical results indicate a fast convergence in the stiffness and damping parameters identification in less than 0.06 s for both considered operation conditions. It is important to mention that the convergence time of the identifier depends mainly on the sample time used in numerical simulations. An important characteristic of the proposed algebraic identifiers is that the unbalance parameters (magnitude and phase) are not needed for their development and implementation because only the vibratory response of the system at the bearings' location and the adjacent nodes is required. As a first approach we have proved the proposed identifiers in rotor-bearing system models with constant rotordynamic coefficients. However, as a future work, the proposed identifiers can be used to numerically and experimentally determine rotordynamic coefficients, which are a function of the system rotational velocity as in the case of pressurized bearings, by adapting the identifier method to estimate non-constant functions. This is possible because in the mathematical model used for the identifiers' development, only the effects (stiffness and damping) of the supports are considered without taking into account the nature of the bearings.

Author Contributions: Conceptualization, J.G.M.-L., M.A.-M. and S.J.L.-D.; methodology, J.G.M.-L., E.B. and J.C.-O.; software, J.G.M.-L. and S.J.L.-D.; validation, J.G.M.-L., S.J.L.-D. and L.A.B.-T.; formal analysis, J.G.M.-L., M.A.-M. and R.T.-H.; investigation, J.G.M.-L., S.J.L.-D. and L.A.B.-T.; resources, J.G.M.-L., E.B., M.A.-M., R.T.-H. and J.C.-O.; data curation, J.G.M.-L., S.J.L.-D. and L.A.B.-T.; writing—original draft preparation, J.G.M.-L., E.B. and M.A.-M.; writing—review and editing, J.G.M.-L., E.B. and M.A.-M.; supervision, E.B. and M.A.-M.; project administration, J.G.M.-L.; funding acquisition, J.G.M.-L. and M.A.-M. All authors have read and agreed to the published version of the manuscript.

Funding: This research received no external funding.

Data Availability Statement: The datasets generated and supporting the findings in the article are obtainable from the corresponding author upon reasonable request.

Acknowledgments: The APC was funded by PRODEP-SEP, Mexico.

Conflicts of Interest: The authors declare no conflict of interest.

Appendix A

Shape functions for the beam finite element.

$$
\begin{aligned}
N_1(y) &= \left[1 - \frac{3y^2}{L^2} + \frac{2y^3}{L^3}; -y + \frac{2y^2}{L} - \frac{y^3}{L^2}; \frac{3y^2}{L^2} - \frac{2y^3}{L^3}; \frac{y^2}{L} - \frac{y^3}{L^2}\right] \\
N_2(y) &= \left[1 - \frac{3y^2}{L^2} + \frac{2y^3}{L^3}; y - \frac{2y^2}{L} + \frac{y^3}{L^2}; \frac{3y^2}{L^2} - \frac{2y^3}{L^3}; -\frac{y^2}{L} + \frac{y^3}{L^2}\right]
\end{aligned}
\tag{A1}
$$

Expressions for matrices in Equation (3) are

$$[M_T] = \frac{\rho S L}{420} \begin{bmatrix} 156 & 0 & 0 & -22L & 54 & 0 & 0 & 13L \\ 0 & 156 & 22L & 0 & 0 & 54 & -13L & 0 \\ 0 & 22L & 4L^2 & 0 & 0 & 13L & -3L^2 & 0 \\ -22L & 0 & 0 & 4L^2 & -13L & 0 & 0 & -3L^2 \\ 54 & 0 & 0 & -13L & 156 & 0 & 0 & 22L \\ 0 & 54 & 13L & 0 & 0 & 156 & -22L & 0 \\ 0 & -13L & -3L^2 & 0 & 0 & -22L & 4L^2 & 0 \\ 13L & 0 & 0 & -3L^2 & 22L & 0 & 0 & 4L^2 \end{bmatrix} \tag{A2}$$

$$[M_R] = \frac{\rho I}{30L} \begin{bmatrix} 36 & 0 & 0 & -3L & -36 & 0 & 0 & 3L \\ 0 & 36 & 3L & 0 & 0 & -36 & 3L & 0 \\ 0 & 3L & 4L^2 & 0 & 0 & 3L & -L^2 & 0 \\ -3L & 0 & 0 & 4L^2 & 3L & 0 & 0 & -L^2 \\ -36 & 0 & 0 & 3L & 36 & 0 & 0 & 3L \\ 0 & -36 & 3L & 0 & 0 & 36 & -3L & 0 \\ 0 & -3L & -L^2 & 0 & 0 & -3L & 4L^2 & 0 \\ -3L & 0 & 0 & -L^2 & 3L & 0 & 0 & 4L^2 \end{bmatrix} \tag{A3}$$

$$[C_1] = \begin{bmatrix} c_{xx} & c_{xz} & 0 & 0 & 0 & 0 & 0 & 0 \\ c_{zx} & c_{zz} & 0 & 0 & 0 & 0 & 0 & 0 \\ 0 & 0 & 0 & 0 & 0 & 0 & 0 & 0 \\ 0 & 0 & 0 & 0 & 0 & 0 & 0 & 0 \\ 0 & 0 & 0 & 0 & 0 & 0 & 0 & 0 \\ 0 & 0 & 0 & 0 & 0 & 0 & 0 & 0 \\ 0 & 0 & 0 & 0 & 0 & 0 & 0 & 0 \\ 0 & 0 & 0 & 0 & 0 & 0 & 0 & 0 \end{bmatrix} \quad or \quad \begin{bmatrix} 0 & 0 & 0 & 0 & 0 & 0 & 0 & 0 \\ 0 & 0 & 0 & 0 & 0 & 0 & 0 & 0 \\ 0 & 0 & 0 & 0 & 0 & 0 & 0 & 0 \\ 0 & 0 & 0 & 0 & 0 & 0 & 0 & 0 \\ 0 & 0 & 0 & 0 & c_{xx} & c_{xz} & 0 & 0 \\ 0 & 0 & 0 & 0 & c_{zx} & c_{zz} & 0 & 0 \\ 0 & 0 & 0 & 0 & 0 & 0 & 0 & 0 \\ 0 & 0 & 0 & 0 & 0 & 0 & 0 & 0 \end{bmatrix} \tag{A4}$$

$$[C_2] = \frac{\rho I}{15L} \begin{bmatrix} 0 & -36 & -3L & 0 & 0 & 36 & -3L & 0 \\ 36 & 0 & 0 & -3L & -36 & 0 & 0 & -3L \\ 3L & 0 & 0 & -4L^2 & -3L & 0 & 0 & L^2 \\ 0 & 3L & 4L^2 & 0 & 0 & -3L & -L^2 & 0 \\ 0 & 36 & 3L & 0 & 0 & -36 & 3L & 0 \\ -36 & 0 & 0 & 3L & 36 & 0 & 0 & 3L \\ 3L & 0 & 0 & L^2 & -3L & 0 & 0 & 4L^2 \\ 0 & 3L & -L^2 & 0 & 0 & -3L & 4L^2 & 0 \end{bmatrix} \tag{A5}$$

$$[K_1] = \begin{bmatrix} k_{xx} & k_{xz} & 0 & 0 & 0 & 0 & 0 & 0 \\ k_{zx} & k_{zz} & 0 & 0 & 0 & 0 & 0 & 0 \\ 0 & 0 & 0 & 0 & 0 & 0 & 0 & 0 \\ 0 & 0 & 0 & 0 & 0 & 0 & 0 & 0 \\ 0 & 0 & 0 & 0 & 0 & 0 & 0 & 0 \\ 0 & 0 & 0 & 0 & 0 & 0 & 0 & 0 \\ 0 & 0 & 0 & 0 & 0 & 0 & 0 & 0 \\ 0 & 0 & 0 & 0 & 0 & 0 & 0 & 0 \end{bmatrix} \quad or \quad \begin{bmatrix} 0 & 0 & 0 & 0 & 0 & 0 & 0 & 0 \\ 0 & 0 & 0 & 0 & 0 & 0 & 0 & 0 \\ 0 & 0 & 0 & 0 & 0 & 0 & 0 & 0 \\ 0 & 0 & 0 & 0 & 0 & 0 & 0 & 0 \\ 0 & 0 & 0 & 0 & k_{xx} & k_{xz} & 0 & 0 \\ 0 & 0 & 0 & 0 & k_{zx} & k_{zz} & 0 & 0 \\ 0 & 0 & 0 & 0 & 0 & 0 & 0 & 0 \\ 0 & 0 & 0 & 0 & 0 & 0 & 0 & 0 \end{bmatrix} \tag{A6}$$

$$[K_2] = A \begin{bmatrix} 12 & 0 & 0 & -6L & -12 & 0 & 0 & -6L \\ 0 & 12 & 6L & 0 & 0 & -12 & 6L & 0 \\ 0 & 6L & (4+a)L^2 & 0 & 0 & -6L & (2-a)L^2 & 0 \\ -6L & 0 & 0 & (4+a)L^2 & 6L & 0 & 0 & (2-a)L^2 \\ -12 & 0 & 0 & 6L & 12 & 0 & 0 & 6L \\ 0 & -12 & -6L & 0 & 0 & 12 & -6L & 0 \\ 0 & 6L & (2-a)L^2 & 0 & 0 & -6L & (4+a)L^2 & 0 \\ -6L & 0 & 0 & (2-a)L^2 & 6L & 0 & 0 & (4+a)L^2 \end{bmatrix} \quad (A7)$$

$$[K_3] = \frac{\rho I}{15L} \begin{bmatrix} 0 & -36 & -3L & 0 & 0 & 36 & -3L & 0 \\ 0 & 0 & 0 & 0 & 0 & 0 & 0 & 0 \\ 0 & 0 & 0 & 0 & 0 & 0 & 0 & 0 \\ 0 & 3L & 4L^2 & 0 & 0 & -3L & -L^2 & 0 \\ 0 & 36 & 3L & 0 & 0 & -36 & 3L & 0 \\ 0 & 0 & 0 & 0 & 0 & 0 & 0 & 0 \\ 0 & 0 & 0 & 0 & 0 & 0 & 0 & 0 \\ 0 & 3L & -L^2 & 0 & 0 & -3L & 4L^2 & 0 \end{bmatrix} \quad (A8)$$

with $A = EI/((1+a)L^3)$ and $a = 12EI/(GSL^2)$, where E is the Young modulus of the shaft material, I is the moment of inertia of the shaft transversal section, a is the shear factor, S is the cross-sectional area of the shaft, L is the element length, G and ρ are the shear modulus and the density of the shaft material, respectively.

References

1. Koutromanos, I. *Fundamentals of Finite Element Analysis: Linear Finite Element Analysis*, 1st ed.; John Wiley & Sons: New York, NY, USA, 2018.
2. Mendoza-Larios, J.G.; Barredo, E.; Colín, J.; Blanco-Ortega, A.; Arias-Montiel, M.; Mayén, J. Computational platform for the analysis and simulation of rotor-bearing systems of multiple degrees of freedom. *Rev. Int. Metod. Numer.* **2020**, *36*, 1–10.
3. Shen, Z.; Chouvion, B.; Thouverez, F.; Beley, A. Enhanced 3D solid finite element formulation for rotor dynamics simulation. *Finite Elem. Anal. Des.* **2021**, *195*, 103584. [CrossRef]
4. Tiwari, R. *Rotor System: Analysis and Identification*, 1st ed.; CRC Press: Boca Raton, FL, USA, 2018.
5. Matsushita, O.; Tanaka, M.; Kobayashi, M.; Keogh, P.; Kanki, H. *Vibrations of Rotating Machinery. Volume 2. Advanced Rotordynamics: Applications of Analysis, Troubleshooting and Diagnosis*, 1st ed.; Springer Japan KK: Tokyo, Japan, 2019.
6. Breńkacz, L. *Bearing Dynamic Coefficients in Rotordynamics. Computation Methods and Practical Applications*, 1st ed.; Wiley-ASME: New York, NY, USA, 2021.
7. Narendiranath Babu, T.; Manvel Raj, T.; Lakshmanan, T. A review on application of dynamic parameters of journal bearing for vibration and condition monitoring. *J. Mech.* **2015**, *31*, 391–416. [CrossRef]
8. Tiwari, R.; Chougale, A. Identification of bearing dynamic parameters and unbalance state in a flexible rotor system fully levitated on active magnetic bearings. *Mechatronics* **2014**, *24*, 274–286. [CrossRef]
9. Xu, Y.; Zhou, J.; Jin, C.; Guo, Q. Identification of the dynamic parameters of active magnetic bearings based on the transfer matrix model updating method. *J. Mech. Sci. Technol.* **2016**, *30*, 2971–2979. [CrossRef]
10. Mao, W.; Han, X.; Liu, G.; Liu, J. Bearing dynamic parameters identification of a flexible rotor-bearing system based on transfer matrix method. *Inverse Probl. Sci. Eng.* **2016**, *24*, 372–392. [CrossRef]
11. Yao, J.; Liu, L.; Yang, F.; Scarpa, F.; Gao, J. Identification and optimization of unbalance parameters in rotor-bearing systems. *J. Sound Vib.* **2018**, *431*, 54–69. [CrossRef]
12. Colín-Ocampo, J.; Guitérrez-Wing, E.S.; Ramírez-Moroyoqui, F.J.; Abúndez-Pliego, A.; Blanco-Ortega, A.; Mayén, J. A novel methodology for the angular position identification of the unbalance force on asymmetric rotors by response polar plot analysis. *Mech. Syst. Signal Process.* **2017**, *95*, 172–186. [CrossRef]
13. Mao, W.; Li, J.; Huang, Z.; Liu, J. Bearing dynamic parameters identification for sliding bearing-rotor system with uncertainty. *Inverse Probl. Sci. Eng.* **2018**, *26*, 1094–1108. [CrossRef]
14. Chen, C.; Jing, J.; Cong, J.; Dai, Z. Identification of dynamic coefficients in circular journals bearings from unbalance response and complementary equations. *Proc. Inst. Mech. Eng. Part J. Eng. Tribol.* **2019**, *233*, 1016–1028. [CrossRef]
15. Rajasekhara, R.M.; Srinivas, J. An optimized bearing parameter identification approach from vibration response spectra. *J. Vibroeng.* **2019**, *21*, 1519–1532.
16. Wang, A.; Yao, W.; He, K.; Meng, G.; Cheng, X.; Yang, J. Analytical modelling and numerical experiment for simultaneous identification of unbalance and rolling-bearing coefficients of the continuous single-disc and single-span rotor-bearing system with Rayleigh beam model. *Mech. Syst. Signal Process.* **2019**, *116*, 322–346. [CrossRef]

17. Wang, X.; Li, H.; Meng, G. Rotordynamic coefficients of a controllable magnetorheological fluid lubricated floating ring bearing. *Tribol. Int.* **2017**, *114*, 1–14. [CrossRef]

18. Kang, Y.; Shi, Z.; Zhang, H.; Zhen, D.; Gu, F. A novel method for the dynamic coefficients identification of journal bearings using Kalman filter. *Sensors* **2020**, *20*, 565. [CrossRef] [PubMed]

19. Chen, Y.; Yang, R.; Sugita, N.; Mao, J.; Shinshi, T. Identification of bearing dynamic parameters and unbalanced forces in a flexible rotor system supported by oil-film bearings and active magnetic devices. *Actuators* **2021**, *10*, 216. [CrossRef]

20. Taherkhani, Z.; Ahmadian, H. Stochastic model updating of rotor support parameters using Bayesian approach. *Mech. Syst. Signal Process.* **2021**, *158*, 107702. [CrossRef]

21. Brito, G.C., Jr.; Machado, R.D.; Neto, A.C.; Kimura, L.Y. A method for the experimental estimation of direct and cross-coupled dynamic coefficients of tilting-pad journal bearings of vertical hydro-generators. *Struct. Health Monit.* **2021**, 1–17. [CrossRef]

22. San Andrés, L.; Jeung, S.H. Orbit-model force coefficients for fluid film bearings: A step beyond linearization. *ASME J. Eng. Gas Turbines Power* **2015**, *132*, 022502.

23. Yang, L.H.; Wang, W.M.; Zhao, S.Q.; Sun, Y.H.; Yu, L. A new nonlinear dynamic analysis method of rotor system supported by oil-film journal bearings. *Appl. Math. Model.* **2014**, *38*, 5239–5255. [CrossRef]

24. Dyk, S.; Smolík, L.; Rendl, J. Predictive capability of various linearization approaches for floating-ring bearings in nonlinear dynamics of turbochargers. *Mech. Mach. Theory* **2020**, *149*, 103843. [CrossRef]

25. Nelles, O. *Nonlinear System Identification. From Classical Approaches to Neural Networks, Fuzzy Models, and Gaussian Processes*, 1st ed.; Springer Nature: Cham, Switzerland, 2020.

26. Tangirala, A.K. *Principles of System Identification: Theory and Practice*, 1st ed.; CRC Press: Boca Raton, FL, USA, 2015.

27. Aster, R.C.; Borchers, B.; Thurber, C.H. *Parameter Estimation and Inverse Problems*, 3rd ed.; Elsevier: Cambridge, MA, USA, 2019.

28. Arias-Montiel, M.; Beltrán-Carbajal, F.; Silva-Navarro, G. On-line algebraic identification of eccentricity parameters in active rotor-bearing systems. *Int. J. Mech. Sci.* **2014**, *85*, 152–159. [CrossRef]

29. Sira-Ramírez, H.; García-Rodríguez, C.; Cortés-Romero, J.; Luviano-Juárez, A. *Algebraic Identification and Estimation Methods in Feedback Control Systems*, 1st ed.; John Wiley & Sons: West Sussex, UK, 2014.

30. Trujillo-Franco, L.G.; Abundis-Fong, H.F.; Campos-Amezcua, R.; Gomez-Martinez, R.; Martinez-Perez, A.I.; Campos-Amezcua, A. Single output and algebraic modal parameters identification of a wind turbine blade: Experimental results. *Appl. Sci.* **2021**, *11*, 3016. [CrossRef]

31. Silva-Navarro, G.; Beltrán-Carbajal, F.; Trujillo-Franco, L.G.; Peza-Solís, J.F.; García-Pérez, O.A. Online estimation techniques for natural and excitation frequencies on MDOF vibrating mechanical systems. *Actuators* **2021**, *10*, 41. [CrossRef]

32. Trujillo-Franco, L.G.; Silva-Navarro, G.; Beltrán-Carbajal, F.; Campos-Mercado, E.; Abundis-Fong, H.F. On-line modal parameter identification applied to linear and nonlinear vibration absorbers. *Actuators* **2020**, *9*, 119. [CrossRef]

33. Beltrán-Carbajal, F.; Silva-Navarro, G. Generalized nonlinear stiffness identification on controlled mechanical vibrating systems. *Asian J. Control* **2019**, *21*, 1281–1292. [CrossRef]

34. Beltran-Carbajal, F.; Silva-Navarro, G.; Trujillo-Franco, L.G. On-line parametric estimation of damped multiple frequency oscillations. *Electr. Power Syst. Res.* **2018**, *154*, 423–432. [CrossRef]

35. Beltrán-Carbajal, F.; Silva-Navarro, G.; Trujillo-Franco, L.G. A sequential algebraic parametric identification approach for nonlinear vibrating mechanical systems. *Asian J. Control* **2017**, *19*, 1–11. [CrossRef]

36. Mendoza-Larios, J.G.; Colín-Ocampo, J.; Blanco-Ortega, A.; Abúndez-Pliego, A.; Gutiérrez-Wing, E.S. Automatic balancing of a rotor-bearing system: On-line algebraic identifier for a rotordynamic balancing system. *Rev. Ibearoam. Autom. Inform. Ind.* **2016**, *13*, 281–292.

37. Colín-Ocampo, J.; Mendoza-Larios, J.G.; Blanco-Ortega, A.; Abúndez-Pliego, A.; Gutiérrez-Wing, E.S. Unbalance determination in rotor-bearing systems at constant velocity: Algebraic identification method. *Ing. Mec. Tecnol. Des.* **2016**, *5*, 385–394. (In Spanish)

38. Beltrán-Carbajal, F.; Silva-Navarro, G. On the algebraic parameter identification of vibrating mechanical systems. *Int. J. Mech. Sci.* **2015**, *92*, 178–186. [CrossRef]

39. Beltrán-Carbajal, F.; Silva-Navarro, G.; Arias-Montiel, M. Active unbalance control of rotor systems using on-line algebraic identification methods. *Asian J. Control* **2013**, *15*, 1627–1637. [CrossRef]

40. Lalanne, M.; Ferraris, G. *Rotordynamics Prediction in Engineering*, 2nd ed.; John Wiley & Sons: West Sussex, UK, 1998.

41. Kharab, A.; Guenther, R.B. *An Introduction to Numerical Methods. A MATLAB Approach*; CRC Pess: Boca Raton, FL, USA, 2019.

MDPI

St. Alban-Anlage 66

4052 Basel

Switzerland

Tel. +41 61 683 77 34

Fax +41 61 302 89 18

www.mdpi.com

Mathematics Editorial Office

E-mail: mathematics@mdpi.com

www.mdpi.com/journal/mathematics